Predictive Analytics in Healthcare, Volume 1

Transforming the future of medicine

IPEM–IOP Series in Physics and Engineering in Medicine and Biology

About the Series

The Series in Physics and Engineering in Medicine and Biology will allow IPEM to enhance its mission to 'advance physics and engineering applied to medicine and biology for the public good'.

This series focuses on key areas including, but not limited to:

- clinical engineering
- diagnostic radiology
- informatics and computing
- magnetic resonance imaging
- nuclear medicine
- physiological measurement
- radiation protection
- radiotherapy
- rehabilitation engineering
- ultrasound and non-ionizing radiation.

A number of IPEM–IOP titles are being published as part of the EUTEMPE Network Series for Medical Physics Experts.

Predictive Analytics in Healthcare, Volume 1

Transforming the future of medicine

Edited by
Vinithasree Subbhuraam
Cyrcadia Health Inc., Houston, TX, USA

IOP Publishing, Bristol, UK

ISBN 978-0-7503-2312-3 (ebook)
ISBN 978-0-7503-2310-9 (print)
ISBN 978-0-7503-2313-0 (myPrint)
ISBN 978-0-7503-2311-6 (mobi)

DOI 10.1088/978-0-7503-2312-3

Version: 20211201

IOP ebooks

British Library Cataloguing-in-Publication Data: A catalogue record for this book is available from the British Library.

Published by IOP Publishing, wholly owned by The Institute of Physics, London

IOP Publishing, Temple Circus, Temple Way, Bristol, BS1 6HG, UK

US Office: IOP Publishing, Inc., 190 North Independence Mall West, Suite 601, Philadelphia, PA 19106, USA

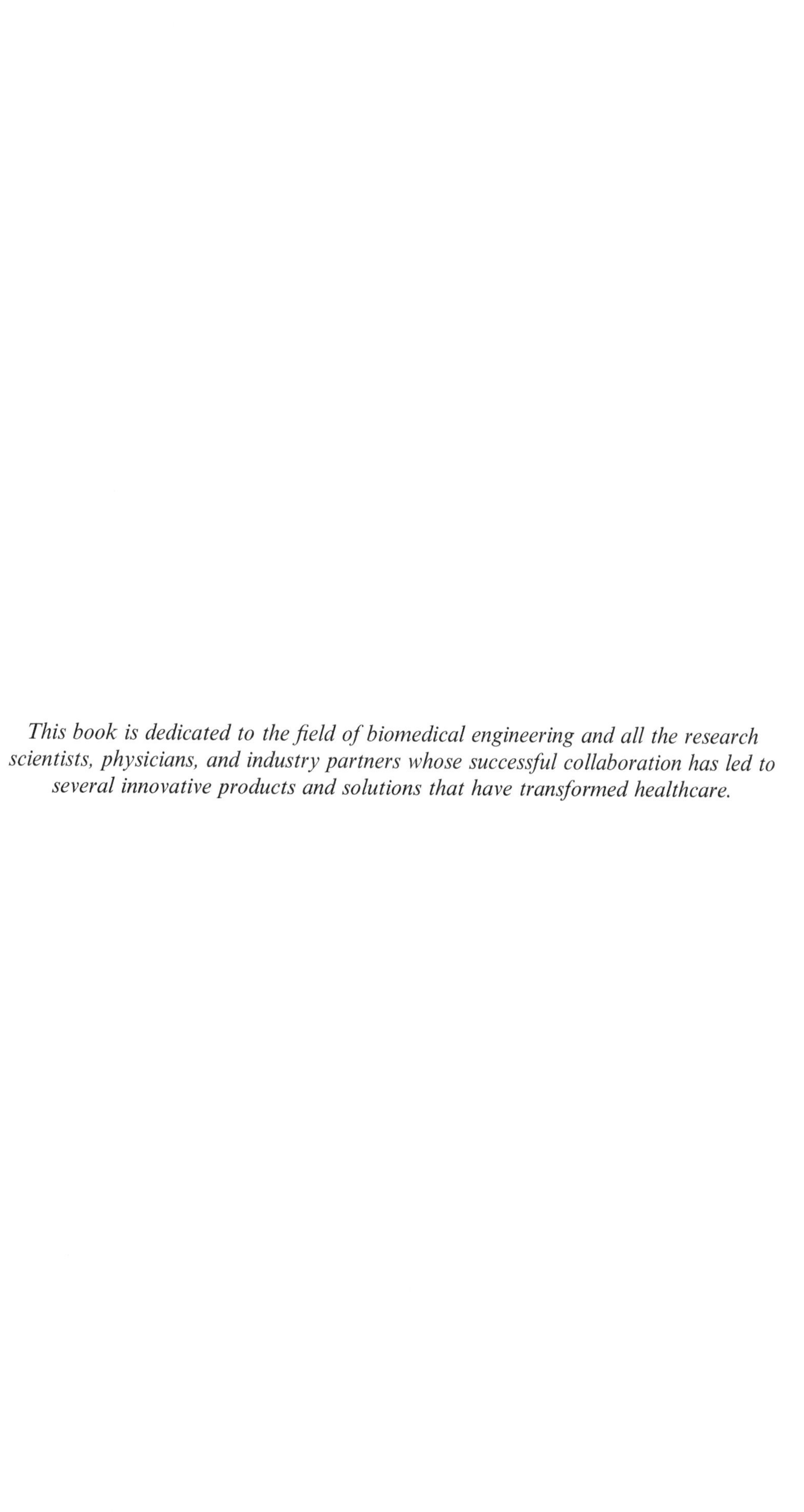

This book is dedicated to the field of biomedical engineering and all the research scientists, physicians, and industry partners whose successful collaboration has led to several innovative products and solutions that have transformed healthcare.

Contents

Preface

Predictive analytics (PA) is a branch of advanced analytics that uses several techniques such as statistics, modeling, data mining, and artificial intelligence to analyze past and current real-time data to gain fast and accurate insights into several aspects of healthcare.

The purpose of this book is to present to the reader details on where and how PA can be employed to improve healthcare. In order to achieve this purpose this book:

- presents an overview of PA for physicians, medical students, biomedical engineers, and data scientists in the healthcare domain
- reviews the current applications of analytics in several healthcare disciplines, so that readers specializing in that field can have a comprehensive overview of all methodologies in place
- enables readers to identify what new applications are needed to advance the use of analytics in their field.

Acknowledgement

I would like to sincerely thank Professor Dr Dhanjoo Ghista for inviting and providing me this opportunity to edit this volume on predictive analytics in healthcare. My heartfelt and profound thanks are due to my supervisors, Professor Dr Eddie Ng Yin Kwee, Nanyang Technological University, Singapore, and Professor Dr Rajendra Acharya, Ngee Ann, Singapore. All three of them have been wonderful mentors and friends to me as I embarked and continued my journey in the field of biomedical engineering. I still remember vividly the conversations I had with them at Nanyang Technological University that laid the foundation for my passion in this field. I thank you all for those precious times.

I also want to express my gratitude to my colleagues at Cyrcadia. They have provided me with immense opportunities, support, and encouragement to learn and grow as a biomedical data scientist and produce a life-changing breast health monitoring solution.

I also wish to extend my sincere thanks to all the reviewers of this book and their valuable suggestions. My thanks are also due to all the esteemed independent authors and my co-authors for finding the time within their busy schedules to contribute their valuable research to this volume, in particular during a pandemic.

Thanks are also due to Jessica Fricchione, Poppy Emerson, Sarah Armstrong, Michael Slaughter, and Phoebe Hooper from IOP Publishing, UK, for their guidance on several areas during the preparation of the manuscript. I am deeply grateful for their patience and understanding as I worked on this volume during the pandemic.

Finally, I would like to express my heartfelt appreciation to my family, in particular my sweet seven-year-old, and to my dear friends, for their wonderful encouragement throughout the completion of this volume and for supporting my passion for biomedical data science.

Editor biography

Vinithasree Subbhuraam

Dr Vinithasree Subbhuraam has over 15 years of experience in biomedical data science and has utilized predictive analytics for designing clinical decision support systems to detect diseases such as carotid atherosclerosis, fatty liver, diabetes, epilepsy, and cancers in the thyroid, breast, ovaries, and prostate. She has MS and PhD degrees in Biomedical Engineering from Nanyang Technological University, Singapore, and a Bachelor's degree in Electronics and Communication Engineering from PSG College of Technology, Coimbatore, India. She is currently the Principal Data Scientist at Cyrcadia. She is also the Associate Editor for *Computer Methods and Programs in Biomedicine* and the Editor (North America) for the *Journal of Medical Imaging and Health Informatics*. She has published over 90 journal papers, conference articles, and book chapters and is a Senior Member of IEEE, and a Full Member of Sigma Xi. She is also an invited reviewer for several international journals such as *IEEE Transactions on Biomedical Engineering*, *IEEE Transactions on Information Technology in Biomedicine*, *Computers in Biology and Medicine*, *Journal of Mechanics in Medicine and Biology*, *Neural Computing and Applications*, *Artificial Intelligence in Medicine*, and *Artificial Intelligence Review*. Her research interests include predictive analytics, machine learning, the Internet of Things, wearable technologies, medical data mining, and clinical decision support systems.

List of contributors

Seifedine Kadry
Faculty of Applied Computing and Technology, Noroff University College, Kristiansand 4608, Norway

Dyuti Kumar
St George's University, School of Medicine, St George's, Grenada, West Indies

Islamiyyat Olatinwo
University of Hertfordshire, UK

Debabrata Panigrahi
Rehabilitation and Regenerative Medicine Lab, Department of Biotechnology and Medical Engineering, National Institute of Technology, Rourkela 769008, India

Vinithasree Subbhuraam
Cyrcadia Health, USA; Cyrcadia Asia, Hong Kong

David Taniar
Faculty of Information Technology, Monash University, Clayton, Victoria 3800, Australia

Rajinikanth Venkatesan
Department of Electronics and Instrumentation Engineering, St Joseph's College of Engineering, Off Old Mamallapuram Road, Chennai, Tamil Nadu 600119, India

Chapter 1

Predictive analytics in healthcare

Vinithasree Subbhuraam

Predictive analytics (PA) is a discipline that uses several techniques such as statistics, modeling, data mining, and artificial intelligence to evaluate past and current real-time data to discover meaningful and actionable patterns that help in making predictions about the future. In healthcare, PA is helpful at every step in a patient's journey, including wellness, preventative care, screening, diagnosis, treatment, and symptom management. In this chapter, an overview of PA is first provided. A brief description of the general framework used to develop predictive models is then given, followed by PA applications in healthcare. Finally, critical challenges in the adoption of PA in healthcare are discussed.

1.1 Introduction

Technological innovation has accelerated the development of digital health systems and services such as mobile health (mHealth), health information technology, wearable devices, telehealth and telemedicine, and personalized medicine. These technologies are designed to empower users to make informed decisions about the prevention, early detection, diagnosis, treatment, and management of several acute and chronic health conditions outside traditional care settings. A critical component of these digital health solutions is prediction, which is key to most decision-making processes. Prediction is generally the process of capturing missing information by evaluating past data, the available data right now, and what is intended to be achieved in the future. It is an art of efficiently combining hindsight, insight, and foresight to discover meaningful and actionable patterns hidden in large amounts of data and use them to achieve valuable outcomes. In healthcare, crunching all the big data will never improve patient care unless actionable insights are determined. The following are three primary reasons to gain actionable insights using predictive analytics in healthcare:

1. *To improve the quality of patient care* by improving clinical decisions using evidence-based medicine.

doi:10.1088/978-0-7503-2312-3ch1

2. *To reduce healthcare costs* by detecting inefficiencies in treatment procedures, forecasting patients' disease risk, etc.
3. *To reduce readmissions and improve hospital operations.*

1.1.1 The role of artificial intelligence in predictive analytics

PA can be performed using simple and complex statistical approaches that can be fine-tuned by humans. However, the technological advancements in the field of artificial intelligence (AI) and medical data mining have led to the use of AI to develop predictive models that use concepts such as machine learning, deep learning, expert systems, natural language processing, etc, to automatically, objectively, and accurately predict more granular and specific outcomes. AI applications in healthcare can make a more significant impact now than before because of the presence of three critical factors: (i) growing interest in addressing the issues associated with the traditional healthcare delivery model, (ii) increased use of networked smart devices in our society, and (iii) increased adaptation to the convenience of at-home services, such as remote health monitoring and telemedicine.

Since society has now started to embrace new health-centric approaches enabled by advances in AI, healthcare organizations are looking for such systems to lower costs and deliver better outcomes. As a result, they have begun to invest heavily in collecting and storing big data from sources such as customer service call logs, clinical documentation, satisfaction surveys, and patient-generated health data from the Internet of Things (IoT). The majority of healthcare organizations still have a great deal of work to effectively and efficiently apply analytics to convert the big data into genuinely actionable clinical intelligence.

Previously, most organizations used descriptive analytics to describe events that happened in the past by using data from sources such as clinical documentation, claims data, patient surveys, lab tests, etc. While this form of analytics can be valuable for clinical and operational management, it does not include forecasting or trending. Predictive analytics, as highlighted earlier, is the ability to extrapolate the course of future events from descriptive data. AI empowers predictive analytics to be faster, smarter, and more actionable than ever before.

1.1.2 The focus of this book

New technical skills and expertise are required to understand how to use AI and PA in healthcare. Several books, publications, and courses online and offline provide comprehensive coverage of the technical implementation side of AI. However, there are not many books that present an application-oriented treatment of analytics in healthcare. Such books should help readers identify where AI can be effectively used in healthcare. For example, in managing a specific disease, the questions to be asked and answered are:

- What type of data should be collected—medical, social, personal, population health-related, genomics—to predict disease risk?

- How can PA and AI be used to predict the changes in lifestyle or symptom management that can help the patient to mitigate the disease's risk of progression?
- Can a specific screening application (app) be developed for the user to keep track of his/her health?
- Once the disease is diagnosed, what are the different ways AI and PA can provide a more objective and definitive diagnosis or help in improving the management and treatment of the condition?
- What are the existing applications available in the market, and what is currently in the research and development phase?

This book aims to provide comprehensive information about where and how PA can be and is currently applied in several different healthcare domains so that the reader can obtain answers to the above questions in their field of interest. The focus in the chapters will be placed on reviewing and analyzing the current and future analytics applications in several healthcare disciplines, which can, later on, contribute to technical implementation. This book does not intend to describe the techniques and technologies needed in healthcare computing and analytics. Instead, it is more application-oriented, and most chapters in this book are written to be accessible to the nontechnical reader.

Section 1.2 of this chapter will introduce the fundamental concepts and principles of AI and PA and describe a generic framework to build predictive models. Section 1.3 will set the premise for the rest of the book by briefly describing PA applications in the healthcare sector.

1.2 The predictive analytics framework

The term 'big data' refers to large amounts of data that are difficult to manage using traditional analytical means and software. Healthcare is a multi-dimensional system that is rich in data. Effective analysis of these data streams to build predictive models is possible by using fast and cost-efficient high-end computational resources and software systematically along with AI algorithms. AI is a field of computer science that aims to imitate human intelligence in computers. This imitation is achieved through iterative, complex design coordination, usually at a speed and scale that surpass human capabilities (Stead 2018). The introduction of AI, bringing together gigantic data streams from diverse specialties, has climaxed in the rise of big data with framework integration over the IoT and cloud computing systems (Ganasegeran and Abdulrahman 2020). AI has been acknowledged as the most powerful analytical tool because it has a framework of computing that permits machines to act or respond to input with a cognitive capacity almost similar to humans (Silver *et al* 2017).

From the healthcare point of view, AI can utilize modern algorithms to learn patterns from an expansive volume of healthcare information (demographics, socioeconomics, medical reports, electronic recordings from medical gadgets, physical examinations, and clinical research facility and images) and subsequently use the acquired knowledge to help clinical practice while using its learning and self-correcting capacities to improve its prediction precision based on feedbacks (Jiang *et al* 2017). AI combined with digital health solutions has been shown to play a critical role in transforming healthcare (Arora 2020).

1.2.1 Concepts of AI

The AI algorithms are generally based on one or more of the following concepts:

- *Machine learning (ML)*: ML is a subfield of AI that consists of algorithms that learn information from past data. The two most commonly used approaches are supervised learning and unsupervised learning. Supervised learning algorithms use input training data and labeled target output data to learn the relationship between the input data and output target. Unsupervised learning algorithms learn patterns in large datasets.
- *Deep learning (DL)*: DL is a subset of ML that employs neural systems and is designed to mimic how a human brain can perform object detection, speech recognition and translation, and decision making. The unique aspect that sets DL apart from ML is learning structured/unstructured and labeled/unlabeled data without much human supervision.
- *Expert systems (ESs)*: ESs are algorithms that try to emulate the decision-making process in humans. They contain a knowledge base, which is a database of medical knowledge, and a rule-based inference engine, which uses the knowledge base to generate predictions. Rule-based expert systems are more understandable to the end-user. However, these systems generally struggle when the database is large. Therefore they are slowly being replaced by ML and DL models.
- *Natural language processing (NLP)*: NLP techniques use algorithms to identify the keywords and phrases in written content and analyze them to generate useful information.

1.2.2 Framework

A predictive model aims to learn associations between the input data attributes and the target outcome so that the learned associations could be used to predict the actual outcome for a new patient. Figure 1.1 shows a general framework used to build and validate a predictive model. Each step will be described briefly in this section.

Step 1. Data collection

Data is the new gold. In healthcare, big data includes data from standard clinical trials and also real-world data (RWD). RWD includes observational data from

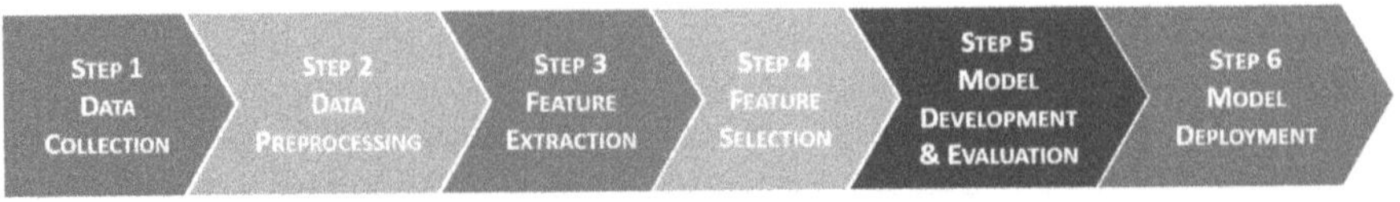

Figure 1.1. General steps for building a predictive model.

electronic health records (EHR), electronic medical records (EMR), clinical examination notes, medical imaging (ultrasound, computed tomography (CT), magnetic resonance imaging (MRI), mammography, positron emission photography (PET), microscopy, etc), inpatient health monitoring sensors (such as electroneurogram (ENG), electromyogram (EMG), electrocardiogram (ECG), electroencephalogram (EEG), phonocardiogram (PCG), etc), patient-generated digital health data from mobile applications and wearable sensors, laboratory results such as genetic markers, hormone levels, complete blood count, etc, 'omics' data (genomics, proteomics, metabolomics, epigenomics, etc), billing and insurance systems, biomedical literature, public health surveillance systems, and social media resources (figure 1.2).

Applying PA to any of these data types will help uncover hidden patterns, correlations, and causations that are generally not evident. For example, medical images are varied, complex, contain irregular shapes, and have noisy values. Interpretations of these images by radiologists may be highly subjective—meaning interpretations vary from person to person based on prior experience. Using machine learning will make the interpretations more objective, repetitive, and reproducible. Further, integration of existing patient data with their genomic, transcriptomic, proteomic, and metabolomic data could provide a deeper understanding of the patient's individual profile. This understanding can help physicians design a personalized healthcare plan for the management and treatment of diseases.

It is apparent that analytics algorithms are highly dependent on the quality of the data inputs to obtain actionable outputs, hence the saying 'garbage in, garbage out'. Each type of data source has different challenges during the data acquisition process. In general, to ensure the quality of data, it is essential to ensure: (i) *data completeness*—e.g. ensure that there are fewer missing values recorded from sensors, a smaller number of missing attributes in EHRs, EMRs, etc, (ii) *data consistency*—the data do not have discrepancies in medical codes or attribute names, (iii) *data relevance*—collect

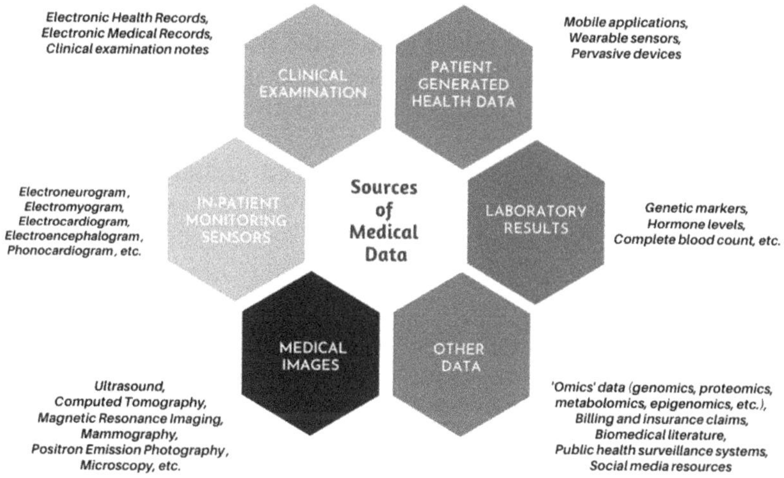

Figure 1.2. Sources of medical data.

only the data that are relevant to the project at hand, and therefore domain knowledge is essential, and (iv) *data privacy*—ensure that the data collected conforms to the Health Insurance Portability and Accountability Act (HIPAA) (Tariq and Hackert 2021).

Step 2. Data pre-processing

Data collection is an expensive and time-consuming task in the healthcare field. For some projects, extensive clinical trials are required to acquire the necessary data in standardized settings. When data are obtained, they are often not in a form that is suitable for further analysis. Data might be collected from varied structured (e.g. EMRs, EHRs) and unstructured (e.g. radiology images, physician notes) sources that need to be appropriately integrated. Despite efforts to ensure data completeness and consistency in the collection phase, data could still have missing entries and outliers. Data pre-processing consists of the steps taken for data cleaning and data integration. It is a critical step in the PA framework which helps immensely in improving the prediction outcome (Duggal *et al* 2016). Alshdaifat *et al* (2021) discuss the impact of data pre-processing steps such as normalization techniques and handling missing values in improving the prediction performance. Reddy and Aggarwal (2015) have compiled detailed chapters on how various pre-processing methods can be applied to different healthcare datasets.

Step 3. Feature extraction

A feature is an individual measurable property or characteristic of the data that could provide more information for prediction than just the raw data. For example, to process radiological images, radiomics features based on image histogram, color, shape, and texture that quantify the variations of the pixel values are extracted and used (van Timmeren *et al* 2020). Other features include simple statistical features such as mean and median, and non-linear features such as entropy. Pre-processed data can also be transformed using mathematical transformations such as the Fourier transform, wavelet transform, principal component analysis (PCA), independent component analysis (ICA), etc, before further modeling (Alturki *et al* 2020).

Step 4. Feature selection

The main idea of feature selection is to choose a subset of features and eliminate the irrelevant ones from the dataset. Obtaining a smaller set of representative features and retaining the optimal salient characteristics of the data decreases the processing time and leads to more compactness of the models learned, better generalization, and comprehensibility of the mined results. There are two basic feature subset selection techniques—wrapper methods, in which the features are selected using the prediction algorithm, and filter methods, in which the selection of features is independent of the prediction model used (Chen *et al* 2020).

Step 5. Model development and evaluation

There are broadly five types of predictive models:

1. *Classification model*: For example, in a problem where a computer-aided detection model has to be built to classify suspicious breast nodules in a mammogram into benign and malignant, the features extracted from the mammograms are the inputs, and the corresponding biopsy results (benign/malignant) are the target labels. A classification model is built to associate

these inputs and the related output labels. The trained model can then predict the category of nodules in new patient mammograms.

2. *Clustering model*: A clustering model groups data/features into separate smaller clusters based on similarity in the features. An example is the clustering model built by Liao *et al* (2016) to identify cost change pattern clusters in healthcare claims datasets belonging to patients with end-stage renal disease who had initiated hemodialysis.
3. *Forecast model*: The forecast model is one of the most widely developed predictive models, particularly in hospital management. These models usually estimate numeric values based on historical data. For example, such models can be designed to forecast optimal staff scheduling based on patient arrival times. Another example is where a multivariate forecasting model was developed to predict the COVID-19 hospital census based on local infection incidence (Turk *et al* 2021). Such models have been critical during the COVID-19 pandemic to plan for adequate staffing and resource allocation.
4. *Outliers model*: An outlier model helps to make predictions based on anomalous data entries in a dataset. One approach is to identify unusual clinical actions in EMRs that could correspond to medical errors. Such outlier-based alerting models could help clinicians determine if a medical error was made or not and if it needs attention (Hauskrecht *et al* 2016).
5. *Time series model*: In the case of a time series model, the predictions for a future time are based on the analysis of data points captured during the current and past periods. For example, to help patients better manage their expenses related to medication costs, Kaushik *et al* (2020) have proposed a time-series-based forecasting model to predict future medication costs based on past purchase patterns over a period of time.

A predictive model is the best combination of selected features and algorithms whose predicted results for a cohort of patients have the maximum correlation to the actual results of the same cohort. In this step, any one or a combination of the above predictive models is built by training the models to associate the predictive features with the target outcome. For this purpose, generally, the original dataset is resampled to form training and test datasets. The training dataset is used for the model building process, and the built model is evaluated using the new unseen test dataset. Two resampling techniques are commonly used to determine the model. One is called the *hold-out technique*, wherein the entire training dataset is used to build the model, and the test dataset is used to evaluate the model. The other method is called the *k-fold cross-validation technique*. Here, the available samples in the training dataset are divided into k approximately equally sized disjoint subsets (stratified by the number of classes—e.g. benign/malignant, high-risk/low-risk, etc). $(k - 1)$ subsets are used for training, and the remaining subset is used for validation to obtain the performance measures. The measures obtained using the validation subset are used to select the best of all the k predictive models. The chosen model is finally evaluated using the reserved test set. The goal of either approach is to estimate the prediction error expected for future use of the model fit to the entire dataset.

The following are some of the most commonly used measures to estimate the prediction performance:

- *Accuracy* refers to the amount of agreement between the predictive model predictions and the actual target results.
- *Sensitivity/recall* refers to the ability of the test to identify those patients with the disease correctly.
- *Specificity* refers to the ability of the test to identify those patients without the disease correctly.
- *Precision/positive predictive value (PPV)* answers the question: 'How likely it is that this patient has the disease given that the test result is positive?'
- *Negative predictive value (NPV)* answers the question: 'How likely it is that this patient does not have the disease given that the test result is negative?'
- *F_1 score* is the weighted mean of precision and recall that indicates a balance between the two measures.

Step 5 is repeated using several predictive algorithms until the best prediction model is determined. Interested readers are referred to the article by Raschka (2018), where other unbiased robust methodologies for model selection and evaluation are outlined in detail. For more information on the algorithms, the reader can refer to the book by Reddy and Aggarwal (2015), wherein several basic and advanced analytical techniques for building predictive algorithms for various healthcare problems are described.

Step 6. Model deployment

Figure 1.3 summarizes the framework used for predictive model development (left) and model deployment to test new real-world datasets (right). The model development includes the steps described in steps 1–5. Predictive features are first extracted and selected from pre-processed clinical data that are relevant to the problem. The features are used as inputs to different predictive algorithms to determine the best model that predicts the future outcome accurately. Once this model is selected, it can be deployed in a healthcare setting (e.g. clinical decision

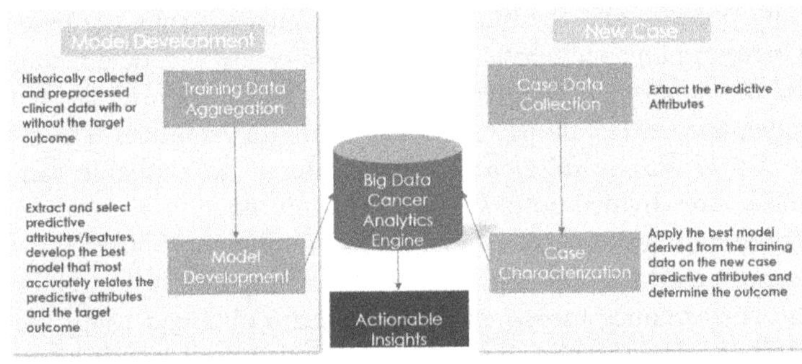

Figure 1.3. Framework for the predictive model development process and analysis of new patient data.

support systems installed in hospital computers, predictive models deployed on the Cloud to be accessed by mobile applications, etc).

When new patient data are received (the right-hand side of figure 1.3), they are first pre-processed to deal with missing values or outliers. The predictive features determined as the best during the model development phase are calculated from this patient's dataset. These features are then fed to the predictive model to determine the outcome for this patient.

Chapters 6 and 7 of this volume provide an idea of how predictive analytics solutions are developed and tested for two specific healthcare conditions using the framework described above. Ultrasound has been used widely to evaluate fetal growth, and several studies have determined that the detection and assessment of fetal head circumference (FHC) are critical to predicting fetal growth. Chapter 6 details a step-by-step computer-assisted image processing procedure to extract FHC. Chapter 7 describes a machine learning framework-based classification model to categorize health fundus images and images that show age-related macular degeneration (AMD).

1.3 Applications

Major theoretical breakthroughs in AI and PA have finally yielded practical applications in healthcare that are poised to change our lives (Bohr and Memarzadeh 2020, Davenport and Kalakota 2019). The following sections describe the six keys areas that will be the focus of this two-volume book.

1.3.1 Public health surveillance

Public health surveillance is defined as the systematic collection, analysis, and interpretation of public health data to gain actionable insights useful for the planning and execution of solutions that improve public health. Predictive analytics and AI can result in a pivotal change in how public health datasets are analyzed (Chiolero and Buckeridge 2020, Zeng *et al* 2021). For example, Google search results data and phone location information have been used to predict restaurants that are causing food-borne illness (Sadilek *et al* 2018). Demographics, biomarkers, and sociomarkers have been used to develop a risk model to identify pediatric asthma patients at risk of hospital revisits (Shin *et al* 2018). In the event of epidemics and pandemics such as COVID-19, mobile phone and social media surveillance has been used extensively to track health seeking behaviors, measure public and private sector facility utilization, and in disease surveillance, behavior monitoring, and public health communication and education (Grantz *et al* 2020, Sun *et al* 2020). Chapter 2 of this volume provides an overview of how predictive analytics has been used extensively for public health surveillance applications such as infectious disease surveillance, environmental health surveillance, and pharmacovigilance.

1.3.2 Women's health

Until just over a couple of decades ago, health research was more focused on the male body. Women were mostly excluded from clinical trials to avoid the impact of hormones and pregnancies on the trial results. However, in the past few years, start-ups and many technology companies have shifted their focus to develop products to manage women's healthcare needs, leading to the rise of female technology (FemTech) solutions. Initial solutions were more focused on reproductive health (menstrual, fertility, and pregnancy care). However, the COVID-19 pandemic has exposed the lack of solutions for managing other conditions such as endometriosis, polycystic ovarian syndrome, thyroid issues, etc. To enable better accessibility to women living in all countries, both in rural and urban locations, several of these solutions are digital-health-based. Many solutions are AI-based algorithms (e.g. breast health monitoring (Cyrcadia 2021), cervical cancer screening (MobileODT 2021), fertility window prediction (Ava Women 2021), etc). Chapter 3 reviews several FemTech solutions in reproductive and non-reproductive women's health issues, emphasizing the use of PA in these solutions.

1.3.3 Pervasive healthcare

Pervasive healthcare is the concept of providing healthcare to anyone, at any time, and anywhere. PA and AI are increasingly being used to power many health-related apps and mobile monitoring devices that utilize sensors (Greco *et al* 2020). Examples of pervasive healthcare include: (i) digital health and IoT applications for the prevention and early screening of various chronic and acute diseases, (ii) tele-medicine services that utilize remote patient monitoring using wearable sensors, (iii) mobile applications that aid in preventative wellness, (iv) smart homes and hospitals embedded with pervasive computing technologies, (v) healthcare monitoring and alert systems for senior care, in particular for people suffering from neurological and cognitive decline, (vi) monitoring systems that support people in palliative care and their caregivers, etc. Chapter 4 is dedicated to providing a detailed review of how various medical specialties have adopted telehealth services and how PA has become an integral part of many of these solutions. Chapter 5 provides an overview of how pervasive healthcare applications can help patients with neurological conditions such as epilepsy, Alzheimer's disease, and Parkinson's disease.

1.3.4 Precision medicine

Precision medicine is a personalized healthcare model that uses computational science, genetics, biology, and social factors to better understand and prevent disease, promote wellness, and develop customized treatment options that allow patients to improve their health and wellness. It has been known that some medicines work for some specific groups of people better than others. With the progressive research on the impact of clinical, molecular, and genomic factors on diseases, it is becoming increasingly possible to develop more personalized and population cohort-based approaches for the detection, management, and treatment of diseases. PA and AI techniques can significantly improve the speed and efficiency of analyzing vast amounts of heterogeneous clinical, multi-omics, and epidemiological datasets (Ahmed 2020). Volume 2 of this book includes a chapter on PA in precision health.

1.3.5 Hospital management

Predictive analytics can help in improving the operational efficiency of healthcare organizations in several ways. The use of PA for efficient hospital management will be the focus of a chapter in volume 2. Predictive models can help analyze patient risk factors and determine their risk for readmissions, enabling efficient resource allocation. Accurate detection of risk predicting factors can also help physicians address the factors to reduce the risk of readmission. PA can also be used to detect early signs of patient deterioration in intensive care units (ICUs) and other wards using automatic early warning systems and biosensor-based remote monitoring. A few other use case scenarios in hospitals include predicting which patients are likely to skip appointments (Ding *et al* 2018), predicting and analyzing patient flow patterns, detecting suicide risks from EMRs (Simon *et al* 2018), and managing hospital supply chains and anticipating equipment maintenance needs.

1.3.6 Chronic disease management

Prediction and prevention go hand-in-hand in chronic disease detection and management. PA can be effectively employed to identify individuals with elevated risks of developing chronic conditions as early in the disease's progression as possible and thereby help avoid long-term health problems that are costly and difficult to treat. Volume 2 will present information in the following areas for the early detection, diagnosis, and management of chronic diseases, such as cardiovascular diseases, cancer, diabetes, and chronic lung diseases such as asthma and chronic obstructive pulmonary disease:

- Predictive analytics techniques and applications that have been developed for the screening, diagnosis, treatment, and management of the disease.
- Commercial products based on PA for the disease and their user guidelines.
- The impact of these technologies and products on physicians and patients.
- Challenges in the implementation and use of these applications.
- Future directions—what data are required by the industry to improve the current applications? What are the new or improved applications necessary for the better management of the disease?

1.4 Challenges

AI applications in healthcare are gaining importance, and more healthcare organizations are figuring out how to harness big data and implement the right infrastructure for generating actionable insights from a multitude of data sources. Some providers are still wondering where and how analytics could be employed effectively in healthcare. This book's primary focus is to guide physicians, medical students, hospital administrators, and the industry to identify analytics applications in healthcare.

The applications mentioned in the earlier section are just the tip of the iceberg. There is a massive opportunity for PA and AI to transform healthcare. A quick PubMed search for 'analytics in healthcare' results in close to 20 000 articles.

However, despite this amount of published research, there is a lack of bench-to-bedside clinical applications or even standardized evaluation of the developed technologies. This translation of biomedical data science research to innovative clinical products and services is possible if we address the following few key issues:

1. **Technical**
    a. *Adequate training.* All stakeholders (physicians and nurses, hospital administrators, medical students, biomedical engineers, data scientists, industry partners, to name a few) should have essential competencies in healthcare and medical data mining concepts and principles.
    b. *Model safety and efficacy.* To ensure efficacy, the predictive models should be developed with (i) reliable and valid datasets, (ii) an adequate number of samples, and (iii) robust and repeatable algorithms and analytics frameworks. To ensure safety, the models should be validated with extensive clinical trials and real-world data, and updated regularly with new data.
    c. *Model transparency.* In an ideal world, data and algorithms should be made public to increase patient and physician confidence and trust. However, this type of transparency is not always possible due to intellectual property protection and the risk of cybersecurity issues. In such cases, the developers should be transparent about the type of data used and the bias in training data (gender, socio-economic factors, demographics, etc). They should also attempt to develop more comprehensible models so that there is some level of understanding about why a particular set of features leads to better predictions (London 2019, Kelly *et al* 2019).
2. **Regulatory**
    In the USA, the Food and Drug Administration (FDA) regulates medical devices. Only in April 2019 did the FDA propose a regulatory framework called 'Modifications to AI/ML-based Software as a Medical Device (SaMD)' for public comment (FDA 2019). More efforts are still needed to establish the framework for these SaMD products. Gerke *et al* (2020a) argue that since there is substantial human involvement in the decision-making process that includes SaMD products, the FDA should not regulate SaMD as products but as overall systems considering the human factor issues. They offer several suggestions on how to regulate AI/ML-based systems in healthcare. Gerke *et al* (2020b) provide an overview of how AI products are regulated in the US and UK and discuss the liability issues surrounding the use of these products. Roberts *et al* (2021) provide a comprehensive outlook on China's AI policy.
3. **Ethical**
    a. *Informed consent.* There is a need to determine when clinicians should deploy principles of informed consent around AI—the type of data used, the nature of the algorithms used, the extent of AI use alongside traditional clinical practice for decision support, patient data privacy and security, etc. For medical apps, it is essential to frame ethically

responsible user agreements that could be quite similar to informed consent documents.

b. *Patient privacy*. It is important to be transparent about how the patient data are used and who owns them. Steps should be taken to ensure that the patient data are not used for purposes other than the intended medical use and discussions with the physician. Care should be taken to protect the patient data from impacting their health insurance premiums, job opportunities, and personal relationships (Gerke *et al* 2019)

1.5 Conclusions

The information included in this book is by no means comprehensive and it is near impossible to describe every single research work and application. However, this book will provide a basic idea of how PA and AI are being used in several healthcare domains and suggests future applications. The editor hopes that the content of this book ignites a passion for biomedical data science in the reader and sparks their interest to research and develop innovative healthcare applications that can be translated from bench to bedside. The applications could be a simple mobile application for the patient, a clinical decision support system for the physician, an efficient EHR management tool for the hospital management team, or a surveillance app for the public health official.

References

Ahmed Z 2020 Practicing precision medicine with intelligently integrative clinical and multi-omics data analysis *Hum Genomics.* **14** 35

Alshdaifat E, Alshdaifat D, Alsarhan A, Hussein F and El-Salhi S M F S 2021 The effect of preprocessing techniques, applied to numeric features, on classification algorithms' performance *Data* **6** 11

Alturki F A, AlSharabi K, Abdurraqeeb A M and Aljalal M 2020 EEG signal analysis for diagnosing neurological disorders using discrete wavelet transform and intelligent techniques *Sensors* **20** 2505

Arora A 2020 Conceptualising artificial intelligence as a digital healthcare innovation: an introductory review *Med. Devices* **13** 223–30

Ava Women 2021 https://avawomen.com/

Bohr A and Memarzadeh K 2020 The rise of artificial intelligence in healthcare applications *Artificial Intelligence in Healthcare* 1st edn ed A Bohr and K Memarzadeh (New York: Academic) ch 2, pp 25–60

Chen C W, Tsai Y H, Chang F R and Lin W C 2020 Ensemble feature selection in medical datasets: combining filter, wrapper, and embedded feature selection results *Expert Syst.* **37** e12553

Chiolero A and Buckeridge D 2020 Glossary for public health surveillance in the age of data science *J. Epidemiol. Community Health* **74** 612–6

Cyrcadia 2021 http://cyrcadiahealth.com/ and https://cyrcadia.asia/

Davenport T and Kalakota R 2019 The potential for artificial intelligence in healthcare *Future Healthc. J.* **6** 94–8

Ding X, Gellad Z F, Mather C 3rd, Barth P, Poon E G, Newman M and Goldstein B A 2018 Designing risk prediction models for ambulatory no-shows across different specialties and clinics *J. Am. Med. Inform. Assoc.* **25** 924–30

Duggal R, Shukla S, Chandra S, Shukla B and Khatri S K 2016 Impact of selected pre-processing techniques on prediction of risk of early readmission for diabetic patients in India *Int. J. Diabetes Dev. Countries* **36** 1–8

FDA 2019 Proposed regulatory framework for modifications to artificial intelligence/machine learning (AI/ML)-based software as a medical device (SaMD) *Report* FDA-2019-N-118 Food and Drug Administration www.fda.gov/files/medical%20devices/published/US-FDA-Artificial-Intelligence-and-Machine-Learning-Discussion-Paper.pdf

Ganasegeran K and Abdulrahman S A 2020 Artificial intelligence applications in tracking health behaviors during disease epidemics *Human Behaviour Analysis Using Intelligent Systems Learning and Analytics in Intelligent Systems* ed D Hemanth vol 6 (Cham: Springer)

Gerke S, Minssen T, Yu H and Cohen I G 2019 Ethical and legal issues of ingestible electronic sensors *Nat. Electron.* **2** 329–34

Gerke S, Babic B, Evgeniou T and Cohen I G 2020a The need for a system view to regulate artificial intelligence/machine learning-based software as medical device *NPJ Digit. Med.* **3** 53

Gerke S, Minssen T and Cohen G 2020b Ethical and legal challenges of artificial intelligence-driven healthcare *Artificial Intelligence in Healthcare* 1st edn ed A Bohr and K Memarzadeh (New York: Academic) ch 12, pp 295–336

Grantz K H *et al* 2020 The use of mobile phone data to inform analysis of COVID-19 pandemic epidemiology *Nat. Commun.* **11** 4961

Greco L, Percannella G, Ritrovato P, Tortorella F and Vento M 2020 Trends in IoT based solutions for health care: moving AI to the edge *Pattern Recognit. Lett.* **135** 346–53

Hauskrecht M, Batal I, Hong C, Nguyen Q, Cooper G F, Visweswaran S and Clermont G 2016 Outlier-based detection of unusual patient-management actions: an ICU study *J. Biomed. Inform.* **64** 211–21

Jiang F *et al* 2017 Artificial intelligence in healthcare: past, present and future *Stroke Vasc. Neurol.* **2** 230–43

Kaushik S *et al* 2020 AI in healthcare: time-series forecasting using statistical, neural, and ensemble architectures *Front. Big Data.* **3** 4

Kelly C J, Karthikesalingam A, Suleyman M, Corrado G and King D 2019 Key challenges for delivering clinical impact with artificial intelligence *BMC Med.* **17** 195

Liao M, Li Y, Kianifard F, Obi E and Arcona S 2016 Cluster analysis and its application to healthcare claims data: a study of end-stage renal disease patients who initiated hemodialysis *BMC Nephrol.* **17** 25

London A J 2019 Artificial intelligence and black-box medical decisions: accuracy versus explainability *Hastings Cent. Rep.* **49** 15–21

MobileODT 2021 https://mobileodt.com/

Raschka S 2018 Model evaluation, model selection, and algorithm selection in machine learning performance estimation: generalization performance vs model selection arXiv: 1811.12808

Reddy C K and Aggarwal C C 2015 *Healthcare Data Analytics* (*Chapman and Hall/CRC Data Mining and Knowledge Discovery Series*) vol 36 (Boca Raton, FL: CRC Press)

Roberts H, Cowls J, Morley J, Taddeo M, Wang V and Floridi L 2021 The Chinese approach to artificial intelligence: an analysis of policy, ethics, and regulation *AI Soc.* **36** 59–77

Sadilek A, Caty S, DiPrete L, Mansour R, Schenk T, Bergtholdt M, Jha A, Ramaswami P and Gabrilovich E 2018 Machine-learned epidemiology: real-time detection of foodborne illness at scale *NPJ Digital Med.* **1** 36

Shin E K, Mahajan R, Akbilgic O and Shaban-Nejad A 2018 Sociomarkers and biomarkers: predictive modeling in identifying pediatric asthma patients at risk of hospital revisits *NPJ Digit. Med.* **1** 50

Silver D *et al* 2017 Mastering the game of Go without human knowledge *Nature* **550** 354–9

Simon G E *et al* 2018 Predicting suicide attempts and suicide deaths following outpatient visits using electronic health records *Am. J. Psych.* **175** 951–60

Stead W W 2018 Clinical implications and challenges of artificial intelligence and deep learning *JÅMÅ* **320** 1107–8

Sun S *et al* 2020 Using smartphones and wearable devices to monitor behavioral changes during COVID-19 *J. Med. Internet Res.* **22** e19992

Tariq R A and Hackert P B 2021 *Patient Confidentiality* (Treasure Island, FL: StatPearls)

Turk P J, Tran T P, Rose G A and McWilliams A 2021 A predictive Internet-based model for COVID-19 hospitalization census *Sci. Rep.* **11** 5106

van Timmeren J E, Cester D, Tanadini-Lang S, Alkadhi H and Baessler B 2020 Radiomics in medical imaging—'how-to' guide and critical reflection *Insights Imaging* **11** 91

Zeng D, Cao Z and Neill D B 2021 Artificial intelligence-enabled public health surveillance—from local detection to global epidemic monitoring and control *Artificial Intelligence in Medicine* (New York: Academic) pp 437–53

Chapter 2

Predictive analytics in public health surveillance

Vinithasree Subbhuraam and Islamiyyat Olatinwo

Predictive analytics (PA) is becoming a trusted methodology for evaluating large and complex public health datasets. It is used to study data to predict the onset and spread of an infectious disease outbreak, the resources needed in various demographics to respond to any crisis, the likelihood of natural disasters, the substance abuse risk in patient cohorts, etc. In this chapter, the basics of a public health surveillance system are first described, followed by a discussion on how PA is harnessed effectively in three critical areas of public health: (i) infectious disease surveillance, (ii) environmental health surveillance, and (iii) pharmacovigilance. Finally, the challenges and future directions of both the technology space and application space are presented.

2.1 Introduction

Thacker and Berkelman defined public health surveillance as the constant systematic collection, analysis, and interpretation of data, thoroughly integrated with the apt dissemination of these data to those liable for preventing and controlling disease and injury (Thacker and Berkelman 1988). The World Health Organization (WHO) also defines public health surveillance as 'An ongoing, systematic collection, analysis and interpretation of health-related data needed for the planning, implementation, and evaluation of public health practice' (WHO 2021a). Public health surveillance covers all fields of public health, ranging from epidemiology, health promotion, health impact assessment, research, environmental and human wellness, and many more. Public health surveillance can be used for the following:

- Measuring patterns and characterizing diseases.
- Estimating the degree and scope of wellbeing issues.
- Identifying epidemics, scourges, wellbeing issues, and changes in health behaviors.
- Monitoring changes in contagious, natural, and environmental agents.
- Pinpointing patients and their contacts for treatment and mediations.

doi:10.1088/978-0-7503-2312-3ch2

- Assessing the viability of programs and control measures.
- Developing hypotheses and fortifying research.

The surveillance should be planned and executed to supply substantial valid data conveniently and economically to decision-makers such as public health personnel, leaders, government, and the public (Nsubuga *et al* 2006). A health surveillance system should be helpful, timely, and characterized by data quality, simplicity, flexibility, stability, sensitivity, acceptability, and representativeness (CDC 2001).

2.2 Steps for building a surveillance system

The steps required to create and maintain a surveillance system include the establishment of goals, development of case definitions, selection of appropriate personnel, acquisition of tools and permission for data collection, analysis, and dissemination, implementation of the surveillance system, and evaluation of surveillance activities (Thacker and Stroup 1998). Once the principal goal of a health surveillance system is established, the next steps include data collection, data analysis, data interpretation, data dissemination, and links to action. The data collection step consists of deciding where and how the information will be collected and the actual collection of information. Once the data collection process is complete, the next major step is analyzing and interpreting the data. The data analysis involves how the information will be examined and what methodology will be used (Smith *et al* 2013). The data interpretation involves how the analyzed data will be translated and decoded. The data dissemination step involves communicating the summary or report of the surveillance to the audience (general public, government, health practitioners, healthcare providers, etc). The link to action monitors the inclinations and patterns in diseases and risk factors (Thacker and Berkelman 1988).

2.2.1 Data collection

2.2.1.1 Types of surveillance systems

An essential step in the data collection process is to decide the type of surveillance system used. The sort of activity that can be taken, when or how regularly that activity must be taken, what data are required to screen the activity, and when or how regularly the data are required decides the type of surveillance or health information system (Foege *et al* 1976). There are different types of public health surveillance systems, as described below.

Active surveillance: This type of surveillance generates the most accurate and timely information as it involves health agencies frequently obtaining information about the condition of wellbeing from the healthcare providers.

Passive surveillance: This type of surveillance system relies on information reported by clinics, hospitals, health departments, laboratories, private practitioners, and other sources. There is no active search for data. The healthcare centers report data regularly. The accuracy cannot be guaranteed as the method depends on the completeness and timeliness of the data provided by these information sources.

Sentinel surveillance: This system collates health issue information from health workers who have been chosen to represent a district or group of people. It can be active or passive surveillance (Birkhead and Maylahn 2000).

Syndromic surveillance: This system focuses on collecting information about symptoms rather than confirmed diseases or physician's diagnoses.

Other types of surveillance include integrated surveillance, categorical surveillance, and behavioral risk factor surveillance systems. With the widespread use of the Internet, social media, and mobile phones, the following types of surveillance are gaining significance:

Event-based surveillance: This type of system collects and analyzes data from various Internet-based search engines, social media, and news outlets. One of the most popular systems is the Global Public Health Intelligence Network (GPHIN) (GPHIN 2020).

Real-time surveillance: This type of system uses web-based tools and tracks search-engine queries to monitor an outbreak in real-time. One example is Google Flu Trends.

Remote-sensing-based surveillance: This type of system uses satellite imagery to monitor changes in environmental parameters such as climate parameters, sea temperatures, soil quality, chlorophyll levels, etc, to predict the outbreak of infectious diseases such as cholera, water-borne diseases, etc. These systems use geospatial technologies such as geographic information systems (GISs) to track the spread of the disease and track public behavior (Saran *et al* 2020).

2.2.1.2 Data sources

A large amount of data can be accumulated using either one or a combination of the above-listed types of surveillance systems chosen based on the goal at hand. Big data refers to the vast amount of data that is increasingly and effortlessly available by digitizing all facets of health, healthcare, and interrelated areas (Murdoch and Detsky 2013). Big data is characterized by volume, velocity, and variety. In the context of public health surveillance, the following are the key sources of data:

Structured data: These are information gathered from electronic health records (EHR), electronic medical records (EMR), and participatory surveillance systems such as crowd mapping and crowdsourcing. The EMR integrates and manages information on each patient gathered in a clinical context, while the EHR encompasses all data related to health throughout life (Murdoch and Detsky 2013).

Semi-structured data: These are information obtained from health monitoring devices.

Unstructured data: These are non-health data with great potential for improving public health. They include data gathered from social media (virtual digital trails), consumption data (real-life digital trails), geographical or spatial data, and physical environment data.

2.2.2 Data analysis and interpretation

Public health surveillance is a necessity for informed planning, execution, and mediation. There is still a need to fill the void between data collection and data interpretation which is the beginning of data usability (Narasimhan *et al* 2004). The challenge lies in analyzing high-dimensional big data, which requires fast and accurate computational techniques and massive storage. The introduction and advancement of artificial intelligence (AI) and PA based techniques have resulted in a pivotal change in how datasets, in general, are analyzed. The following section lists and describes some key applications of PA in public health surveillance.

2.3 Predictive analytics in public health surveillance

As described in chapter 1, predictive analytics can be used to build predictive models to analyze data coming in from public health surveillance systems effectively. Some of the critical applications of predictive analytics in health surveillance are:

- Predicting the onset and spread of an infectious disease outbreak.
- Environmental health surveillance.
- Pharmacovigilance.
- Predicting the probability of a natural disaster.
- Determining the amount and types of vaccines required in various demographics.
- Predicting the likelihood of a biological or chemical attack.
- Providing an estimate of multiple parameters in a clinical setting: health insurance risk profiling, staffing requirements, medical errors, appointments, etc.
- Detecting aberrations, which are recognizable proof of strange incidents or patterns in information, with clinical pertinence (Bi *et al* 2019).

In this section, a few of the critical applications will be described in further detail.

2.3.1 Infectious disease surveillance

The WHO (2021b) defines 'infectious diseases as the diseases that are caused by pathogenic micro-organisms, such as bacteria, viruses, parasites or fungi; the diseases that can be spread, directly or indirectly, from one person to another; zoonotic diseases are infectious diseases of animals that can cause disease when transmitted to humans'. The occurrence of infectious diseases can be affected by several factors, including changes in land use and environmental acts, microbial variations and adaptation, the development of technology and industry, the breakdown of public health measures, demographic and behavioral changes, business and international travel, etc (Institute of Medicine 1992). These factors have increased disease emergence, disease transferability, and host susceptibility. Factors influencing the emergence of disease persist in the twenty-first century. With food-borne diseases and antimicrobial resistance, many chronic diseases have been discovered to be affected by infectious diseases.

Some infectious diseases may result in minor to severe symptoms, while some infections are asymptomatic. Infectious diseases are hazardous and pose severe implications to global health because they disrespect boundaries. Globalization and rapid urbanization have made it easier for the strains of diseases to spread (Agrebi and Larbi 2020). The spread of these diseases differs from organism to organism. Three overlapping periods are critical in infectious diseases: (i) *the incubation period*, which is the period between when the infection takes place and the time for the disease to take root, (ii) *the latent period*, which is the period between the infection taking place and the patient becoming infectious, and (iii) *the period of communicability*, which is the time once the patient becomes infectious or contagious (Cohen 2000).

At the start of the twentieth century, infectious diseases were the primary cause of mortality worldwide. In the United States, three illnesses, namely tuberculosis, pneumonia, and diarrhea, were responsible for 30% of deaths. In the mid-twentieth century, mortality from numerous infectious diseases declined, and the introduction of antimicrobial agents quickened this regression even more. By the end of the twentieth century, deaths from infectious diseases were replaced by death from chronic illnesses such as heart disease, cancer, and stroke in most developed countries (CDC 1994). However, developing countries did not report the same success with infectious diseases. According to CDC, in 1998 (CDC 1998), infectious diseases were responsible for 13 million deaths, almost a quarter of the total mortality globally. At the end of the twentieth century, new disease strains and micro-organisms were detected. Some previously controlled diseases were re-emerging and since then the danger of rising and re-emerging infectious diseases has remained significantly large. Therefore, the intervention capabilities essential for standing up to such pandemics and risks must become more efficient (Agrebi and Larbi 2020).

For decades, disease surveillance has been practiced and it has become an integral approach for identifying developing illnesses, outbreaks, and plagues. Early information about an illness outbreak plays a critical role in increasing the efficiency of interventions. Conventional infection surveillance regularly depends on research facility-based diagnoses that are generally time-consuming, and the study of notifiable illnesses is frequently slow-paced and deficient. In contrast, a computer-based surveillance framework has the potential to essentially hasten the discovery of infection outbreaks (Chen and Zeng 2009). These modern, computer-based surveillance frameworks offer necessary and convenient data to clinics, healthcare providers, health agencies, and non-government and government health authorities. These frameworks use carefully acquired and cleaned large databases, details about infectious disease informatics, and algorithms based on machine learning, deep learning, natural language processing, and high-quality visualization to present and monitor real-time surveillance details. With these recent signs of progress, public health surveillance and observation frameworks are becoming efficient in the real-time recognition of severe sicknesses and the potential risk of bioterrorism, permitting swift public health measures.

The concern about the rate and mode of transmission of infectious diseases has driven healthcare authorities to enhance the means of early detection of people at risk of infectious diseases. In this regard, AI-based predictive analytics techniques can be used effectively in infectious disease surveillance to assess the risk of an outbreak, hasten diagnosis, study and mitigate transmission, determine efficient treatments, find vaccines, study antimicrobial resistance, and track public behavior (figure 2.1). Summaries of a few recent and relevant studies in each area are presented in the following sub-sections.

2.3.1.1 Assessing the risk of outbreaks

Predictive analytics can be used to predict the risk of an outbreak ahead of time. Unlike traditional risk assessment models that use known risk factors, the predictive risk models developed using machine learning can use and learn from known and unknown risk factors and can also be tailored easily to specific demographics or health facilities. Moreover, these models can also be dynamically adjusted for individual patients as their conditions change. Such customizable approaches enable infection prevention and control teams to be proactive rather than reactive and initiate quick and efficient care. This approach can also be used to accurately identify and recruit patients at high risk of a particular infection for clinical trials of novel therapies. Several groups have used this machine learning approach to successfully predict the risk of infections, for example, nosocomial *Clostridioides difficile* infection (Li *et al* 2019, Oh *et al* 2018).

2.3.1.2 Early and rapid diagnosis

It is necessary to accurately and quickly predict/diagnose infectious disease in a patient to start immediate treatment and implement quick contact-tracing measures to control the spread of the disease. Predictive models can be developed for the same. A few examples of studies utilizing AI for early and rapid diagnosis of infectious diseases are presented below:

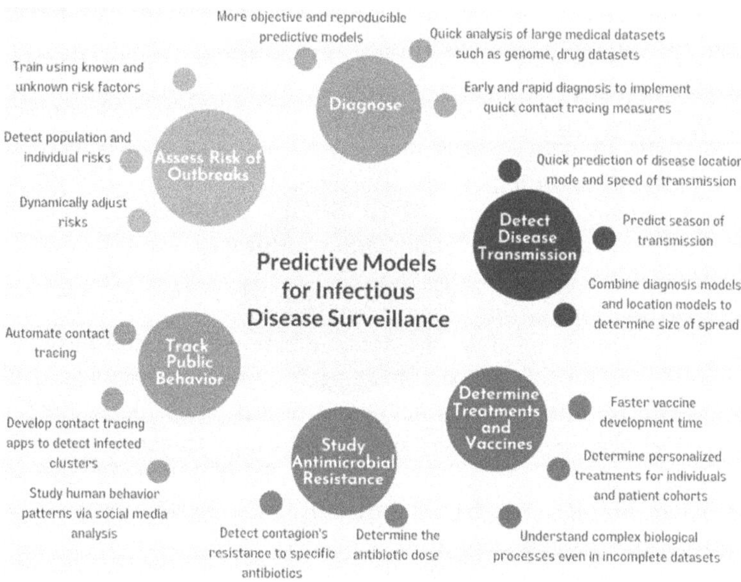

Figure 2.1. The use of predictive models for infectious disease surveillance.

- *H5N1 influenza*: Sun *et al* (2015) developed an infectious disease and fever screening radar system that used Kohonen's self-organizing map neural network and fuzzy clustering methodology to effectively separate higher risk H5N1 influenza patients from low-risk and no-flu groups. Patients' vital signs such as respiration rate (measured by a radar system), heart rate, and facial temperature were used as input to the predictive model. The critical element of this model was that it could screen patients into different groups within ten seconds, enabling rapid diagnosis and subsequent use of control measures.
- *Tuberculosis (TB)*: TB is a bacterial infectious disease caused by *Mycobacterium tuberculosis* and is one of the leading causes of death by an infectious disease globally. AI can enable a more accurate and objective diagnosis of TB through the use of a variety of data sources. The most commonly used data source for TB detection is chest radiographs. Several deep learning-based models have been proposed to diagnose TB in chest radiograph images (Khan *et al* 2020, Lakhani and Sundaram 2017). Other data sources include, but are not limited to, using pathology images (Xiong *et al* 2018) and genomic data to study drug resistance to specific TB gene mutations (Jamal *et al* 2020).
- *Malaria*: Go *et al* (2018) developed a support vector machine-based machine learning model using features extracted from microscopy images to detect unstained red blood cells infected with malaria. The trained model could detect infected cells with an accuracy of 97.50%. The proposed model is quicker and simpler to implement in practice as it does not require staining procedures before testing with the model. Such AI-assisted microscopy techniques have several advantages: more rapid analysis, more objective results without subjective inter-observer variabilities, and the ability to transfer and analyze images in a central reading station in places with no clinical microbiologist on site (Smith *et al* 2018).

Next-generation sequencing (NGS) is a type of DNA sequencing technique that can sequence the whole human genome in a single day. It is widely used to identify pathogens rapidly (Deurenberg *et al* 2017, ECDPC 2016). AI-based techniques enhance and speed up the analysis of NGS data and integrate other healthcare datasets to predict or detect correlations and causations.

2.3.1.3 Studying and mitigating disease transmission

Compared to non-communicable diseases, infectious diseases have the unique characteristic of spreading from person to person. The contagion's virulence, the severity of symptoms, and the associated mortality rate necessitate the quick prediction of the disease's location, size, mode, and transmission speed. Other parameters such as the time of incubation and availability of treatments should also be studied. AI techniques can be utilized efficiently for this purpose to develop mathematical models to understand the transmission of infectious diseases.

For example, Majumdar *et al* (2018) developed a classification framework for Kyasanur forest disease (KFD), a tick-borne viral infectious disease endemic to

South Asia. To prevent an endemic disease from becoming an epidemic, they proposed a fog computing-based health framework that incorporated a novel neural network-based classification algorithm to detect infected patients and a Global Positioning System (GPS) based location alert system to monitor affected patients. This effective combination of a disease-predicting algorithm plus a location alert mechanism enables the quick identification of the size and location of transmission. Subsequent rapid measures can be taken to control or prevent an outbreak.

Once an outbreak has been identified, it is necessary to study the transmission metrics such as seasonality. For example, Mohammed *et al* (2018) used techniques such as the seasonal ARIMA (SARIMA) and neural network auto-regression (SARIMA-NNAR) to forecast pulmonary tuberculosis incidence in Iraq. They were also able to determine that spring and winter were the peak seasons for incidence and transmission. Predicting seasons for peak incidence ahead of time can help optimize the allocation and use of healthcare resources more efficiently.

2.3.1.4 Determining treatments and vaccines

The immune system utilizes multiple cell types to determine the immune response necessary to address any pathogen that enters the human body. In immunological studies, these cell types, molecules, and gene expressions are determined using computational methods, enabling appropriate vaccine development. One key element in this process is to account for biological variability in different populations and patient subgroups, such as pregnant women. Another issue is to understand these complex biological processes in incomplete datasets. Machine learning and deep learning techniques can be effectively employed in this scenario to understand complex biological processes using multiple algorithms on such datasets. For example, Tomic *et al* (2019) proposed a machine learning workflow called SIMON (Sequential Iterative Modeling 'OverNight'), which could be effectively used on incomplete datasets to determine features that correlate to immunity for a particular infectious disease such as the common flu.

Predictive analytics can also drastically cut down the time of the vaccine development process due to the ability to handle complex datasets with increasingly available computational power and resources. These techniques can sort through tens of thousands of sub-components of a virus and predict which of them are most capable of producing an immune response. In a more recent innovation in this space, Baidu developed a LinearFold AI algorithm to predict the secondary nature of the ribo nucleic acid (RNA) sequence of a virus, which has led to the development of messenger RNA vaccines (Huang *et al* 2019).

AI techniques have also been employed widely in the treatment and drug discovery space (Fleming 2018). Shen *et al* (2018) proposed a decision support system that detects infectious disease based on a patient's disease condition and recommends an antibiotic treatment adapted to the patient based on body temperature, infection sites, symptoms, drug interactions, antibacterial spectrum complications, and contradictions.

2.3.1.5 Studying antimicrobial resistance

Sometimes, bacteria and viruses become drug-resistant, and it will become difficult to treat them effectively using existing drugs. Bioinformaticians are using artificial intelligence-based techniques to detect and address this issue of antibiotic/antimicrobial resistance. In one study by Nguyen *et al* (2019) a machine learning model called XGBoost was trained to accurately predict the minimum inhibitory concentrations (MICs) for 15 antibiotics used against non-typhoidal *Salmonella* infection. For training the model, they used sequences of *Salmonella* genomes as the input and MICs as the target output. On a new test genome strain, the trained model would predict the antibiotics that the strain would be resistant to and the relevant dose.

Tackling antibiotic resistance can be enhanced with the Comprehensive Antibiotic Resistance Database (CARD), which provides quality reference data on the molecular basis of antimicrobial resistance (Jia *et al* 2017, Alcock *et al* 2020). CARD is existentially structured and has information about antimicrobial resistant drug groups and mechanisms.

2.3.1.6 Tracking public behavior

Data mining solutions have been implemented in some studies to track public behavior and human behavior patterns during disease outbreaks (Ganasegeran and Abdulrahman 2020). These solutions have helped to identify disease hotspots and enhance timely public health interventions while mitigating transmission. In a study by Alicino *et al* (2015), Google search queries related to Ebola incidence in West Africa were analyzed. As expected, there was a strong correlation between the number of Ebola cases and the Google trend metrics. In weaker correlation areas, the most common causes were the low media coverage and limited availability of digital solutions and the Internet. Another interesting aspect of such analyses is the provision to understand and discover the nature of the queries (panic-induced when there were zero cases in the region yet, information-based to learn about the disease and its symptoms, or queries to understand the personal hygiene measures required to prevent the spread).

Many countries have also encouraged the citizens to install contact-tracing apps to track incidence and control spread. For example, Singapore has an app called TraceTogether to enable efficient contact-tracing. The app has sufficient measures to protect user privacy. It uses short-distance Bluetooth signal exchange between users near each other to register records of these close contacts locally in the user's own devices. When a contact-tracing investigation happens, the users are then required to share this data.

2.3.1.7 Special mention: COVID-19

During the most recent COVID-19 pandemic, PA based techniques have been explored widely for virus surveillance, identifying interventions and potential drug responses, developing vaccines, predicting high-risk populations, and supporting health services through AI-based mobile apps for remote consultation and general public information sharing (Mbunge *et al* 2020).

Screening: During the early stages of COVID-19, it is difficult to detect accurately due to the similarities of symptoms to several other flu-like diseases. Moreover, the most commonly employed polymerase chain reaction (PCR) test has limited sensitivity, and results can take up to 72 h. In such a scenario, AI-based techniques that use other blood test-based lab parameters or image-based data (computed tomography (CT) images, chest x-rays) as sources could prove more reliable, quick, and objective in screening and diagnosis.

In a study by Li *et al* (2020) a deep learning-based predictive/diagnostic model was developed to accurately detect and differentiate COVID-19 from other community-acquired pneumonia and non-pneumonia abnormalities in chest CT images. The model demonstrated a sensitivity and specificity of 90% and 96%, respectively. Ardakani *et al* (2020) also developed a deep learning framework that could differentiate CT images from COVID and pneumonia patients with an accuracy of 86.27%. Studies led by Sun *et al* (2020) and Wu *et al* (2020) applied analytics on clinical lab parameters from patients' bloodwork to understand which parameters were affected by the presence of COVID. This type of analysis enables the implementation of rapid diagnosis AI algorithms that utilize these parameters as features and gives a comprehensive insight into the disease itself. In a similar large-scale study utilizing around 115,394 patients (Soltan *et al* 2021), clinical data generally captured within 1 h of arrival at the hospital were used to train machine learning predictive models to differentiate COVID patients from pre-pandemic controls. The proposed models were able to achieve more than 90% accuracy in comparison with PCR tests.

Contact-tracing: Contact-tracing helps identify the people who came in close contact with an infected person and is key to breaking the chain of virus transmission. Traditional contact-tracing involves an infected person providing information about their recent contacts and activities. This approach's success depends heavily on the contact's recall accuracy and the time taken to notify contacts manually (Braithwaite *et al* 2020). Moreover, the process is time-consuming and resource-intensive. Automation of this approach could address some of these limitations. Several countries have now successfully implemented contact-tracing digitally using apps using technologies such as Bluetooth, GPS, mobile data tracking, etc (Lalmuanawma *et al* 2020). Kricka *et al* (2020) list a few digital contact-tracing tools that are AI-powered. Such apps can also provide insights into trends and associations between digital contact-tracing, tests, and disease outbreaks.

AI techniques can utilize data from social media and news outlets to generate applications such as Healthmap (https://healthmap.org/en/). These applications can also then be used for further contact-tracing by using pattern recognition algorithms. Bluedot (https://bluedot.global/) is another similar AI platform that utilizes air travel data for outbreak surveillance (Fitzpatrick *et al* 2020).

Treatment and vaccines: Because of the capability of AI/PA algorithms, specifically deep learning techniques, to quickly and comprehensively analyze large databases, researchers have employed them to study the efficacy of existing drugs for COVID treatment. A research group from Taiwan found eight drugs to treat COVID from among 80 existing drugs (Ke *et al* 2020). Similarly, a collaborative group from the US and Korea determined that a popular antiretroviral HIV drug called Antazanavir and another drug called Remdisivir could be effective for COVID treatment (Beck *et al* 2020). As indicated earlier, the LinearFold AI algorithm proposed by the team at Baidu (Huang *et al* 2019) has been used successfully to predict the COVID RNA sequence, which has led to the development of messenger RNA vaccines for COVID.

2.3.2 Environmental health surveillance

Several models have been developed to describe the social determinants of health. These models describe how our environments' social, economic, and physical elements interact with our biological factors and behaviors to determine our health status. One of the most widely referred to models is the rainbow model proposed by Dahlgren and Whitehead (1991). In this model, the individuals are placed in the middle with a set of genetic and biological factors that they do not have control over. Surrounding them are the various external influencing layers such as individual lifestyle factors, social and community networks, and other socioeconomic, cultural, and environmental conditions (education, food, water, sanitation, healthcare, housing, work). These encompassing layers of determinants can be adjusted to reduce health inequalities and promote health.

For example, living conditions in a neighborhood, such as the adequate availability of sidewalks and medical care, socioeconomic conditions such as poverty rate, and environmental conditions such as pollutants, can affect the health of the individuals living in that neighborhood. These conditions can continually change over time. Monitoring these changes could predict the effect of adverse neighborhood conditions and improve these social and physical environments to provide good health for all and achieve health equity. Schootman *et al* (2016) reviewed several emerging technologies that can aid in neighborhood condition surveillance. Some common ways to measure these conditions are by analyzing the data acquired by technologies such as Google Street View, webcams, remote sensing, social media, and unmanned aerial vehicles. Crowdsourcing applications are applications that use data from a large group of people to achieve a particular objective. Google Earth (http://earth.google.com/), GPS data received from mobile devices, apps such as HealthMap's Outbreaks Near Me, and several others can revolutionize how data are captured to gain insights about the population in any specific area. Boulos *et al* (2011) provide an excellent review of how crowdsourcing, citizen sensing, and sensor-based technologies are used for environmental health surveillance. They also discuss the trends and critical challenges, such as noise in the data, information overload, privacy, and trust.

In another interesting study, Kelly *et al* (2014) assessed the association between observed physical activity behavior and characteristics of the built environment to understand the healthcare status of the particular location under study. Traditional survey-based methods to understand the physical activity happening in a particular location are limited by the reliability of self-reporting of individuals living in that location. Emerging observational technologies that use high-resolution omnidirectional imagery such as Google Street View and Microsoft Earth can be used to unobtrusively assess physical activity and evaluate the environment (mixed land-use, pedestrian infrastructure, and all non-residential land use). In another virtual systematic social observational study, Google Street View was used to capture and measure neighborhood conditions and their relation to children's antisocial behavior (Odgers *et al* 2012). The study reported Google Street View as a dependable and cost-efficient tool for local neighborhood surveillance. Algorithms based on machine learning, deep learning, natural language processing, image mining, text mining, and pattern recognition can be effectively employed to analyze these large datasets.

Shin *et al* (2018) introduced the concept of sociomarkers as the measurable indicators of a patient's social conditions. They proposed machine learning-based models that use demographics (age, race, gender), zip-code based sociomarkers (insurance type, blight prevalence, broken windows ratio, housing quality, neighborhood inequality, poverty level, presence of trash), biomarkers (length of hospitalization and symptom severity) to detect pediatric asthmatic patients at risk of hospital revisit following an initial visit. The sociomarker-based model and the biomarker-based model achieved 61% and 65% accuracies, respectively.

Another example is the use of predictive analytics tools to combine data from EHR and prescription drug monitoring program (PDMP) data with a patient's social determinants of health to develop a profile for patients at risk of substance abuse and help providers anticipate how their prescription decisions might affect those patients (Tseregounis and Henry 2021).

2.3.3 Pharmacovigilance

Pharmacovigilance (PV) is another aspect of health surveillance that involves the discovery, evaluation, understanding, and deterrence of the unfavorable effects of drugs or any other drug-related issues (Meyboom *et al* 1999). PV is primarily used to monitor the safety and long-term adverse drug reactions (ADR) observed during the use of new drugs that have just entered the market. Between 2008 and 2017, the US Food and Drug Administration (FDA) approved 321 new drugs. During this time the FDA also documented more than 10 million ADR reports, among which 5.8 million were severe and 1.1 million were deadly (Basile *et al* 2019). There is a need for a quick, efficient, and objective way to analyze all these ADR reports and determine correlations and causations. AI and AP can be used effectively for this purpose. AI solutions can be implemented in PV to gather accurate data, manage and sort new data from varied sources, and identify adverse drug events and reactions (Murali *et al* 2019). Basile *et al* (2019) have summarized several recent machine learning and deep learning approaches to assess preclinical drug safety and post-marketing drug surveillance.

2.4 What is next?

Surveillance is a critical topic in public health informatics, with an increasing application of AI approaches to handle large complex datasets efficiently. There is no dearth to the amount of data available from all types of sources in the current digital era, including but not limited to social media, EHRs, EMRs, and geographical and spatial data. Advancements in AI technologies and predictive modeling enable us to analyze substantial public health-related datasets and gather critical information that could potentially help monitor and improve public health (Wiens and Shenoy 2018).

In this chapter, the steps in designing a public health surveillance system were provided. Several key application areas were presented to introduce to the reader how predictive analytics can be used to improve public health. What is presented is just scratching the surface of what has been done and what could be done next. The several references provided in this chapter will lead the reader to discover further knowledge in this space. In summary, a few of the essential advancements necessary in both the technology space and application areas are listed in this section.

2.4.1 Challenges and advancements necessary in the technology space

In a growingly interconnected world, there is immense potential to harness the power of data analytics and AI to perform public health surveillance. Below are a few of the challenges that need to be carefully addressed to implement automatic public health surveillance systems successfully.

1. *Data collection*: There is always an issue of missing data in the healthcare space, primarily because of the healthcare provider's neglect in capturing all required parameters. Missing data are even more of an issue in low- and middle-income countries that lack the digital infrastructure to capture and maintain information. This is compounded by the fact that there is also a lack of well-trained professionals. Global data collection standards should be established and followed to acquire good quality datasets (Kamel Boulos *et al* 2011). Cloud-based digital infrastructure systems could help low-income countries participate in the process (Wahl *et al* 2018). These global datasets should be shared with researchers worldwide so that each group can use the same dataset to develop predictive models. It will then be easier to compare models and detect the best-performing model for a particular dataset.
2. *Data security and privacy*: Data collected by surveillance systems could contain personal information that could be misused. Data anonymization and strict adherence to privacy laws applicable to specific countries should be followed. Moreover, adequate data security measures should be in place. Patient consent is also of paramount importance.
3. *Data integration*: At present, health data can come from various sources: hospital and community EMRs, EHRs, and other data infrastructure from multiple merchants, smartphones, wearables, social media, spatial data from Google Earth, etc. To develop comprehensive and valuable predictive models, the input should be a well-integrated dataset formed from all these sources.
4. *Data analysis*: Several predictive analytic algorithms can behave like 'black boxes', meaning it is difficult to comprehend how the algorithm arrived at a

prediction. Rule-based models are more explanatory as they indicate what parameters were used for prediction, but they are not always easy to develop. In the case of training machine learning models, domain experts who have designed traditional surveillance systems should be consulted to understand what parameters are significant for a particular problem. The data scientist can start from there and add more features to the predictive models. Moreover, crowdsourced data, in particular from social media, tend to be noisy. Experts could help separate the crowd's personal opinion from credible data. Designing a predictive model with a certain level of domain knowledge is the first step to producing comprehensible results.

5. *Model validation*: A model developed using a smaller dataset representative of a specific population might not have the same prediction accuracy in a different population. It is best to have a training dataset that is a mix of several demographics so that the predictive model can learn the representative features that are common or very specific to populations. Further, the prediction accuracy will continue to improve with the increasing use of the model and re-training processes with more incoming datasets. Sometimes the models can provide ambiguous predictions for the same dataset. In such scenarios, an expert in the field should be consulted to take further action to interpret outputs and ensure clinical relevance. Thus, instead of developing fully automated AI solutions, it is best to have AI-assisted solutions called 'human in the loop' AI. This will allow for a better embracing of the technology in the healthcare space where predictions can directly affect patient care. Every little care should be taken to ensure accurate predictions.
6. *Regulatory approval*: Even though there are regulatory bodies such as GDPR that provide guidelines for using and analyzing personal data for a public health aim (Ienca and Vayena 2020), there is still a need for a global regulatory group. This group should decide on common data collection standards, establish guidelines for safe and secure data use, storage and sharing, policies for development and validation of predictive models, regulatory clearance, etc. Decisions taken by this global team should allow for the approved technology to be used worldwide at the same time. A uniform deployment enables global citizens to enjoy the advantages of health surveillance. It will allow us to understand the predictive model's efficiency better and make improvements with global parameters. Policymakers in all countries should also be educated about the advantages, risks, and opportunities of AI-based surveillance systems.

2.4.2 Future directions in the application space

1. *Precision Public Health*: Precision Public Health is a term coined to indicate how external socioeconomic factors can be used along with an individual's genetic and biological markers to improve the individual's health. Health surveillance incorporating social determinants of health can improve

decision-making, suggesting relationships and pathways of health inequalities (Khoury *et al* 2018). Although there are very few research articles on the use of artificial intelligence in public health surveillance, it is an indication of new directions and possible paradigm shift.

2. *Using emerging technologies*: There are several emerging technologies in addition to AI that can work together to enhance disease surveillance, in particular in the presence of an endemic disease, epidemic, or pandemic. Some of them are the Internet of Medical Things (IoMT), virtual reality (VR), and autonomous robots. The IoMT framework can enable remote monitoring of infected patients who are in self-isolation or quarantine facilities and also provide telehealth support for other health issues in the patient. VR technology can simulate an interactive 3D physical reality in a virtual setting. This set-up could be utilized in training healthcare professionals without the need for physical contact. Moreover, mental health support can also be provided by VR (Javaid *et al* 2020). Autonomous robots can collect samples from patients, disinfect and sterilize contaminated areas, and distribute drugs to patients to limit healthcare workers' exposure.

Future work should focus on strengthening these current technologies, including AI, and developing a robust inter-operable intelligent framework that combines the strengths of all these technologies. Such a regulated model could be immediately deployed globally and locally in an infectious disease outbreak to control the spread and save lives.

References

Agrebi S and Larbi A 2020 Use of artificial intelligence in infectious diseases *Artificial Intelligence and Precision Health* (New York: Academic) pp 415–38

Alcock B P *et al* 2020 CARD 2020: antibiotic resistome surveillance with the comprehensive antibiotic resistance database *Nucleic Acids Res.* **48** D517–25

Alicino C *et al* 2015 Assessing Ebola-related web search behavior: insights and implications from an analytical study of Google Trends-based query volumes *Infect. Dis. Poverty* **4** 54

Ardakani A A, Kanafi A R, Acharya U R, Khadem N and Mohammadi A 2020 Application of deep learning technique to manage COVID-19 in routine clinical practice using CT images: results of 10 convolutional neural networks *Comput. Biol. Med.* **121** 103795

Basile A O, Yahi A and Tatonetti N P 2019 Artificial intelligence for drug toxicity and safety *Trends Pharmacol. Sci.* **40** 624–35

Beck B R, Shin B, Choi Y, Park S and Kang K 2020 Predicting commercially available antiviral drugs that may act on the novel coronavirus (SARS-CoV-2) through a drug-target interaction deep learning model *Comput. Struct. Biotechnol. J.* **18** 784–90

Bi Q *et al* 2019 What is machine learning? A primer for the epidemiologist *Am. J. Epidemiol.* **188** 2222–39

Birkhead G S and Maylahn C M 2000 State and local public health surveillance *Principles and Practices of Public Health Surveillance* ed S M Teutsch and R E Churchill vol 270 (New York: Oxford University Press)

Braithwaite I, Callender T, Bullock M and Aldridge R W 2020 Automated and partly automated contact tracing: a systematic review to inform the control of COVID-19 *Lancet Digit. Health* **2** e607–21

CDC 1994 *US Centers for Disease Control Prevention 1994 Fact Book 7* (Atlanta, GA: CDC)

CDC 1998 *Preventing Emerging Infectious Diseases. A Strategy for the 21st Century* (Atlanta, GA: US Department of Health and Human Services)

CDC 2001 Outbreak of Ebola Hemorrhagic Fever, Uganda, August 2000–January 2001 *Morbidity and Mortality Weekly Report* 50 US Centers for Disease Control and Prevention pp 73–7

Chen H and Zeng D 2009 AI for global disease surveillance *IEEE Intell. Syst.* **24** 66–82

Cohen M L 2000 Changing patterns of infectious disease *Nature* **406** 762–7

Dahlgren G and Whitehead M 1991 *Policies and Strategies to Promote Social Equity in Health* (Stockholm: Institute for Future Studies)

Deurenberg R H *et al* 2017 Application of next generation sequencing in clinical microbiology and infection prevention *J. Biotechnol.* **243** 16–24

ECDPC 2016 Expert opinion on whole genome sequencing for public health surveillance *Scientific Advice Report* European Centre for Disease Prevention and Control, Stockholm https://ecdc.europa.eu/sites/default/files/media/en/publications/Publications/whole-genome-sequencing-for-public-health-surveillance.pdf

Fleming N 2018 How artificial intelligence is changing drug discovery *Nature* **557** S55–7

Fitzpatrick F, Doherty A and Lacey G 2020 Using artificial intelligence in infection prevention *Curr. Treat. Options Infect. Dis.* **12** 135–44

Foege W H, Hogan R C and Newton L H 1976 Surveillance projects for selected diseases *Int. J. Epidemiol.* **5** 29–37

Ganasegeran K and Abdulrahman S A 2020 Artificial intelligence applications in tracking health behaviors during disease epidemics *Human Behaviour Analysis Using Intelligent Systems Learning and Analytics in Intelligent Systems* ed D Hemanth vol 6 (Cham: Springer)

Go T, Kim J H, Byeon H and Lee S J 2018 Machine learning-based in-line holographic sensing of unstained malaria-infected red blood cells *J. Biophotonics* **11** e201800101

GPHIN 2020 https://gphin.canada.ca/cepr/aboutgphin-rmispenbref.jsp?language=en_CA

Huang L, Zhang H, Deng D, Zhao K, Liu K, Hendrix D A and Mathews D H 2019 LinearFold: linear-time approximate RNA folding by 5′-to-3′ dynamic programming and beam search *Bioinformatics* **2019** i295–304

Ienca M and Vayena E 2020 On the responsible use of digital data to tackle the COVID-19 pandemic *Nat. Med.* **26** 463–4

Institute of Medicine 1992 *Emerging Infections: Microbial Threats to Health in the United States* (Washington, DC: National Academy Press)

Jamal S *et al* 2020 Artificial intelligence and machine learning based prediction of resistant and susceptible mutations in *Mycobacterium tuberculosis Sci. Rep.* **10** 5487

Javaid M, Haleem A, Vaishya R, Bahl S, Suman R and Vaish A 2020 Industry 4.0 technologies and their applications in fighting COVID-19 pandemic *Diabetes Metab. Syndr.* **14** 419–22

Jia B *et al* 2017 CARD 2017: expansion and model-centric curation of the comprehensive antibiotic resistance database *Nucleic Acids Res.* **45** D566–73

Kamel Boulos M N *et al* 2011 Crowdsourcing, citizen sensing and sensor web technologies for public and environmental health surveillance and crisis management: trends, OGC standards and application examples *Int. J. Health Geogr.* **10** 67

Ke Y-Y *et al* 2020 Artificial intelligence approach fighting COVID-19 with repurposing drugs *Biomed. J.* **43** 355–62

Kelly C, Wilson J S, Schootman M, Clennin M, Baker E A and Miller D K 2014 The built environment predicts observed physical activity *Front. Public Health.* **2** 52

Khan F A, Majidulla A, Tavaziva G, Nazish A, Abidi S K, Benedetti A, Menzies D, Johnston J C, Khan A J and Saeed S 2020 Chest x-ray analysis with deep learning-based software as a triage test for pulmonary tuberculosis: a prospective study of diagnostic accuracy for culture-confirmed disease *Lancet Digit. Heal.* **2** E573–81

Khoury M J, Engelgau M, Chambers D A and Mensah G A 2018 Beyond public health genomics: can big data and predictive analytics deliver precision public health? *Public Health Genomics.* **21** 244–50

Kricka L J *et al* 2020 Artificial intelligence-powered search tools and resources in the fight against COVID-19 *EJIFCC* **31** 106–16

Lakhani P and Sundaram B 2017 Deep learning at chest radiography: automated classification of pulmonary tuberculosis by using convolutional neural networks *Radiology* **284** 574–82

Lalmuanawma S, Hussain J and Chhakchhuak L 2020 Applications of machine learning and artificial intelligence for Covid-19 (SARS-CoV-2) pandemic: a review *Chaos Solitons Fractals* **139** 110059

Li B Y, Oh J, Young V B, Rao K and Wiens J 2019 Using machine learning and the electronic health record to predict complicated *Clostridium difficile* infection *Open Forum Infect. Dis.* **6** ofz186

Li L *et al* 2020 Using artificial intelligence to detect COVID-19 and community-acquired pneumonia based on pulmonary CT: evaluation of the diagnostic accuracy *Radiology* **296** E65–71

Majumdar A, Debnath T, Sood S K and Baishnab K L 2018 Kyasanur forest disease classification framework using novel extremal optimization tuned neural network in fog computing environment *J. Med. Syst.* **42** 187

Mbunge E, Akinnuwesi B, Fashoto S G, Metfula A S and Mashwama P 2020 A critical review of emerging technologies for tackling COVID-19 pandemic *Hum. Behav. Emerg. Technol.* **3** 25–39

Meyboom R H, Egberts A C, Gribnau F W and Hekster Y A 1999 Pharmacovigilance in perspective *Drug Saf.* **21** 429–47

Mohammed S H, Ahmed M M, Al-Mousawi A M and Azeez A 2018 Seasonal behavior and forecasting trends of tuberculosis incidence in Holy Kerbala, Iraq *Int. J. Mycobacteriol.* **7** 361–7

Murali K, Kaur S, Prakash A and Medhi B 2019 Artificial intelligence in pharmacovigilance: practical utility *Indian J. Pharmacol.* **51** 373–6

Murdoch T B and Detsky A S 2013 The inevitable application of big data to health care *JAMA* **309** 1351–2

Narasimhan V *et al* 2004 Responding to the global human resources crisis *Lancet* **363** 1469–72

Nguyen M, Long S W, McDermott P F, Olsen R J, Olson R, Stevens R L, Tyson G H, Zhao S and Davis J J 2019 Using machine learning to predict antimicrobial MICs and associated genomic features for nontyphoidal salmonella *J. Clin. Microbiol.* **57** e01260–18

Nsubuga P *et al* 2006 Public health surveillance: a tool for targeting and monitoring interventions *Disease Control Priorities in Developing Countries* ed D T Jamison *et al* 2nd edn (Washington, DC: The International Bank for Reconstruction and Development/The World Bank)

Odgers C L, Caspi A, Bates C J, Sampson R J and Moffitt T E 2012 Systematic social observation of children's neighborhoods using Google Street View: a reliable and cost-effective method *J. Child Psychol. Psychiatry* **53** 1009–17

Oh J *et al* 2018 A generalizable, data-driven approach to predict daily risk of *Clostridium difficile* infection at two large academic health centers *Infect. Control Hosp. Epidemiol.* **39** 425–33

Saran S, Singh P, Kumar V and Chauhan P 2020 Review of geospatial technology for infectious disease surveillance: use case on COVID-19 *J. Indian Soc. Remote Sens.* **48** 1121–38

Schootman M *et al* 2016 Emerging technologies to measure neighborhood conditions in public health: implications for interventions and next steps *Int. J. Health Geogr.* **15** 20

Shen Y, Yuan K, Chen D, Colloc J, Yang M, Li Y and Lei K 2018 An ontology-driven clinical decision support system (IDDAP) for infectious disease diagnosis and antibiotic prescription *Artif. Intell. Med.* **86** 20–32

Shin E K, Mahajan R, Akbilgic O and Shaban-Nejad A 2018 Sociomarkers and biomarkers: predictive modeling in identifying pediatric asthma patients at risk of hospital revisits *NPJ Digit. Med.* **1** 50

Smith P F *et al* 2013 Blueprint version 2.0: updating public health surveillance for the 21st century *J. Public Health Manag. Pract.* **19** 231–9

Smith K P, Kang A D and Kirby J E 2018 Automated interpretation of blood culture gram stains by use of a deep convolutional neural network *J. Clin. Micro.* **56** e01521–17

Soltan A A S, Kouchaki S, Zhu T, Kiyasseh D, Taylor T, Hussain Z B, Peto T, Brent A J, Eyre D W and Clifton D A 2021 Rapid triage for COVID-19 using routine clinical data for patients attending hospital: development and prospective validation of an artificial intelligence screening test *Lancet Digit. Health.* **3** e78–87

Sun G, Matsui T, Hakozaki Y and Abe S 2015 An infectious disease/fever screening radar system which stratifies higher-risk patients within ten seconds using a neural network and the fuzzy grouping method *J. Infect.* **70** 230–6

Sun L *et al* 2020 Combination of four clinical indicators predicts the severe/critical symptom of patients infected COVID-19 *J. Clin. Virol.* **128** 104431

Thacker S B and Berkelman R L 1988 Public health surveillance in the United States *Epidemiol. Rev.* **10** 164–90

Thacker S B and Stroup D F 1998 Public health surveillance *Applied Epidemiology: Theory to Practice* ed R C Brownson and D B Petitti (New York: Oxford University Press) pp 105–35

Thacker S B, Berkelman R L and Stroup D F 1989 The science of public health surveillance *J. Public Health Policy.* **10** 187–203

Tomic A, Tomic I, Rosenberg-Hasson Y, Dekker C L, Maecker H T and Davis M M 2019 SIMON, an automated machine learning system, reveals immune signatures of influenza vaccine responses *J. Immunol.* **203** 749–59

Tseregounis I E and Henry S G 2021 Assessing opioid overdose risk: a review of clinical prediction models utilizing patient-level data *Transl. Res.* **234** 74–87

Wahl B, Cossy-Gantner A, Germann S and Schwalbe N R 2018 Artificial intelligence (AI) and global health: how can AI contribute to health in resource-poor settings? *BMJ Glob. Health.* **3** e000798

Wiens J and Shenoy E S 2018 Machine learning for healthcare: on the verge of a major shift in healthcare epidemiology *Clin Infect Dis.* **66** 149–53

WHO 2016 Public health surveillance, World Health Organization

WHO 2021a Public health surveillance, World Health Organization https://who.int/immunization/monitoring_surveillance/burden/vpd/en/

WHO 2021b Infectious diseases, World Health Organization http://emro.who.int/health-topics/infectious-diseases/index.html

Wu J *et al* 2020 Rapid and accurate identification of COVID-19 infection through machine learning based on clinical available blood test results *medRxiv* https://doi.org/10.1101/2020.04.02.20051136

Xiong Y, Ba X, Hou A, Zhang K, Chen L and Li T 2018 Automatic detection of *Mycobacterium tuberculosis* using artificial intelligence *J. Thorac. Dis.* **10** 1936–40

IOP Publishing

Predictive Analytics in Healthcare, Volume 1
Transforming the future of medicine
Vinithasree Subbhuraam

Chapter 3

FemTech solutions for advancing women's health

Vinithasree Subbhuraam

FemTech refers to the range of digital health technologies that focus specifically on women's health. A majority of the solutions include wearable devices, smartphone applications, connected medical devices, and digital platforms to address women's health issues. Several of these solutions rely heavily on predictive analytics (PA) to monitor and render timely interventions. In this chapter, the advantages of using PA in the FemTech space are first discussed. A detailed review of various FemTech applications for reproductive health (menstrual health, fertility, menopause, pregnancy, and maternal care) and non-reproductive health (cancers of the breast, ovary, and cervix, chronic and autoimmune conditions, and general wellness) is then provided, along with indications on how analytics has been employed effectively in these applications. Finally, a few general recommendations to improve applications and platforms in the FemTech space are presented.

3.1 Introduction

Women are about 49.5% of the global population (Countrymeters 2021). However, in 2018 only 4.7% of the total healthcare funding in the US went towards developing products and services for women's health (Rock Health 2021). For a long time, researchers were concerned that women's menstrual and hormonal cycles would complicate research experiments. Only in 1994 is the National Institutes of Health mandate the inclusion of women in clinical trials, enabling a better understanding and research of women's health data. Women and men are different in terms of anatomy and metabolisms (Fitzpatrick and Thakor 2019). Therefore, there is a critical need to include women in clinical trials and develop products specific to improving women's health. Gender-specific healthcare services and solutions are the need of the hour.

doi:10.1088/978-0-7503-2312-3ch3

In recent years, the technological field has taken significant steps to address gender disparities giving rise to FemTech (short for 'female technology') solutions. It was a term coined by Ida Tin, the co-founder and CEO of the period tracking app, Clue. It comprises digital tech solutions (software, hardware, wearables, Internet of Medical Things, smartphone applications) to improve women's health in the areas of general health and wellness, reproductive health (menstruation, fertility, pregnancy, prenatal and postnatal care, and sexual health) (figure 3.1), cancers (breast, ovarian, cervical), and other chronic and autoimmune conditions (figure 3.2). Frost and Sullivan project that by 2025, FemTech solutions will have a market size of up to USD 50 billion (Frost and Sullivan 2021).

This chapter first provides a brief overview of how PA and artificial intelligence (AI) solutions are helpful in the FemTech space. It then reviews a few FemTech solutions for managing women's reproductive and non-reproductive health. Many of these solutions are data-driven and use PA, and indications on how analytics has been employed effectively in these applications are also provided.

3.2 PA and AI in the FemTech space

Because of the late inclusion of women in research trials and the lack of prioritization of women's health needs, there is a lack of data specific to women's health. Many FemTech solutions aim to address these data gap issues using wearables for collecting continuous longitudinal data. Several healthcare companies and start-ups in the FemTech space use predictive modeling and AI-based techniques to develop interactive and integrated digital health solutions for women's health. These solutions aim to provide personalized coaching and education for lifestyle changes and ways for early disease risk prediction, detection, and diagnosis, symptom monitoring, and treatment.

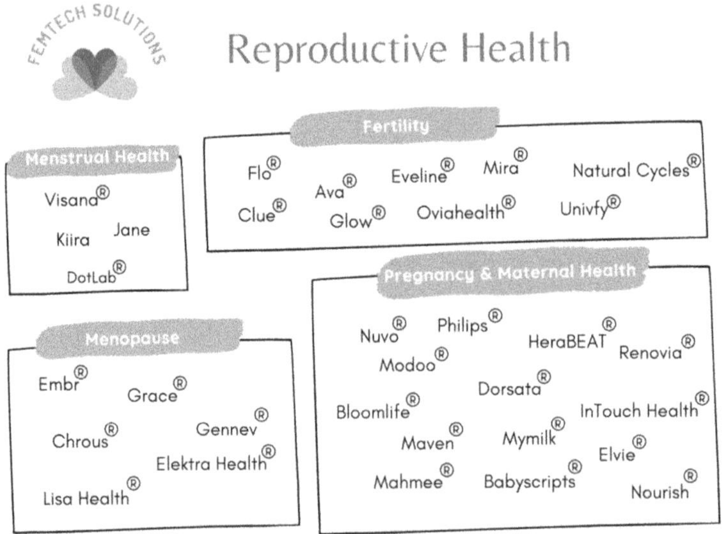

Figure 3.1. A few FemTech solutions in the reproductive health space.

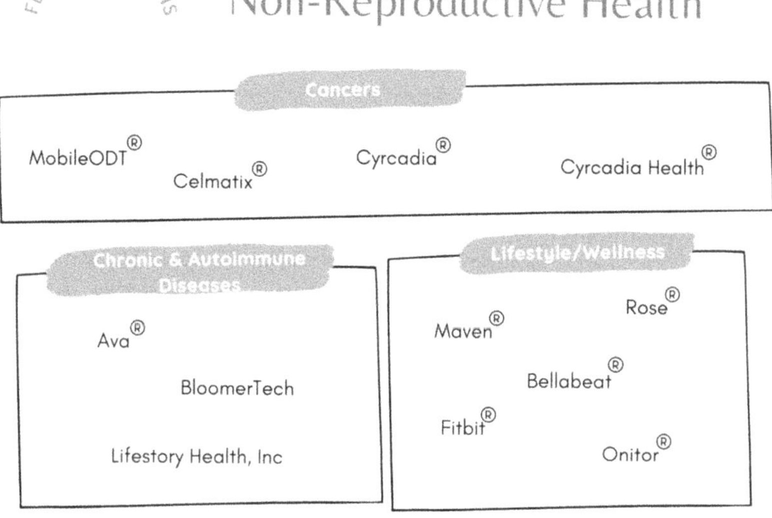

Figure 3.2. A few FemTech solutions in the non-reproductive health space.

Advantages of using PA and AI in the FemTech space include the following:

- *Insightful analysis*: Longitudinal data analyzed using AI could provide more details on physiological changes than data obtained using a single test at a doctor's office. For example, the continuous data collected by period tracking apps such as Clue and Flo from millions of women worldwide could be used to build data models and predictive analytic algorithms to study, understand, and develop solutions around reproductive health. Further, solutions targeted at specific women's health conditions could also provide valuable insight into other indications of general sickness. For example, the Ava fertility tracker is now being studied in a 20 000 women large-scale clinical trial to determine whether the temperature, breathing, and heart rate data collected using the bracelet help to detect the early onset of COVID-19 symptoms.
- *Updated insights*: As life progresses, subtle changes can happen in the data collected, and these subtle changes could have a significant impact on the health condition that is being tracked, for example, using a wearable device. Without adequate updating and ongoing maintenance of analysis algorithms, the usability of data decays over time. Traditional generic app updates or device updates are generally cumbersome and may miss capturing personalized changes occurring in an individual. On the other hand, embedded AI-powered algorithms can iteratively improve over time using newly accrued data. These algorithms are beneficial in FemTech solutions. They can track parameters during various hormonal changes, and provide timely predictions and recommend appropriate interventions to handle the impact of changing hormones.

- *Integrated analysis*: With the technological advancements in algorithms, predictive models and AI-based solutions can now be developed quickly using data from multiple structured and unstructured data sources—such as apps, wearables, connected medical devices, electronic health records, clinical notes, etc. Such an integrated approach allows for the development of personalized solutions customized to each woman's individual physical, biological, and socioeconomic conditions.
- *Address gender and demographic bias*: AI-powered algorithms can quickly learn the characteristics of diverse datasets collected worldwide. This capability is critical for developing accurate solutions that work for the global population and can also be customizable for local groups. Moreover, AI algorithms could also be trained using only data from women to develop targeted solutions for women's health. For example, because of hormonal changes, the symptoms of heart diseases present themselves differently in women. An AI algorithm trained using a dominant male dataset will more likely misdiagnose heart attacks in women than men. This misdiagnosis is because the algorithm would have failed to learn from the unique indicators of a heart attack in women. Thus, targeted female data collection is critical to continue to innovate in the women's health space.
- *Reduced cost and time of diagnosis*: Embedded AI algorithms in wearables and handheld devices may help provide timely and affordable healthcare services to remote and rural areas in developed and developing countries worldwide. These may be of particular benefit in developing countries. For example, in countries that do not have national cancer screening programs, wearable screening devices using AI could help detect cancer early and provide a means for non-invasive detection that would encourage more women to participate.

3.3 FemTech applications for reproductive health

3.3.1 Menstrual health

Eighty-four per cent of women experience pain during their menstrual cycles, and it affects their daily life and work (Grandi *et al* 2012). Visana Health (2021) provides a 12-week digital program based on nutrition, pelvic floor physical therapy, disease education, and cognitive behavioral therapy to help reduce pelvic pain and brain fog symptoms, chronic menstrual pain, etc. Users are provided access to a personal coach, and the platform also offers science-based content to help women understand and manage any condition. Platforms such as Kiira Health (2021) connects young women with an AI-based care coordinator and also provides telemedicine access to a range of women's health providers.

Endometriosis is a painful disorder in which endometrial tissue that lines the inside of the uterus is present in other organs outside the uterus. The condition causes painful menstrual cycles and may also lead to infertility. It affects approximately 10% of reproductive age women globally (WHO 2021b). Biopsies are currently the most definitive way to diagnose the disease (Agarwal *et al* 2019).

Non-invasive tests based on microRNAs are emerging which may detect endometriosis (Moga *et al* 2019). DotLab (Moustafa *et al* 2020, DotLab 2021) has developed a product called DotEndo consisting of a blood test that measures the expression of microRNA biomarkers and statistical algorithms to detect the presence or absence of endometriosis.

NextGen Jane is a data-driven FemTech company planning to offer a woman insight into her reproductive health by analyzing the genomic signals from the endometrial tissue shed during menstruation (NextGen Jane 2021).

3.3.2 Fertility

Fertility-awareness-based methods (FABMs) use scientific information about fertility and each woman's menstrual cycle together to predict the fertile days in the woman's cycle. Traditional FABMs track the basal body temperature (BBT) rhythm during the menstrual cycle. The BBT is usually the lowest before ovulation (Stephenson and Kolka 1999). The fertile window is defined as beginning five days before the day of ovulation and ending on the day of ovulation. While tracking BBT seems to be a straightforward method, its accuracy depends on reliable daily recordings of BBT by the woman, preferably at the same time. Moreover, accurate BBT can also be affected by environmental factors, altered sleep times, time zone changes, etc, and the measurement requires a thermometer sensitive enough to capture small changes in BBT.

Another way of determining the fertile window is to track the rise in a hormone called the luteinizing hormone (LH) in urine samples. Even though LH testing is non-invasive and more accurate than BBT tracking, the purchase of LH test strips over many cycles can prove costly. Moreover, LH test kits only predict half of the fertile window, missing the first few days of the fertile window where women have the highest probability of conception.

To address the limitations of these traditional FABMs, new, mobile phone-based, and wearable-based solutions have been developed to monitor menstruation and fertility (Earle *et al* 2020). Shilaih *et al* (2018) studied the correlation of wrist skin temperature (WST) measured using wearable sensors during sleep and LH tests in detecting ovulation. They confirmed a WST shift over the fertile window. They observed that menstrual-cycle-based changes in WST were not affected by other factors such as alcohol consumption, eating before bedtime, etc. They also noted that since WST is captured continuously during the night, automatic temperature shift detection algorithms embedded in smartphone apps could be used effectively. Maijala *et al* (2019) studied finger skin temperatures recorded by the wearable Oura ring during the night to predict menstruation and ovulation. They also observed that this technique showed the potential for menstrual cycle monitoring.

In another study, Jennings *et al* (2019) evaluated a smartphone app called Dynamic Optimal Timing (DOT) that estimates the fertile window using the woman's start date of the menstrual period. DOT then uses machine learning predictive algorithms to identify days when the woman is at high or low chance of pregnancy by considering and adapting to the user's cycle lengths and variability data.

A prevalent FemTech solution is a period tracker. These trackers help women capture and track period-related data and associated symptoms, empowering them with monthly personal data. The data help women monitor their menstrual cycle, determine ovulation to help in conception or avoid pregnancy, and understand physical and emotional changes associated with their cycle to manage their symptoms better. Some of the popular period tracking apps are listed below:

1. *Flo* (Flo Health 2021): The Flow Period and Ovulation Tracker app captures the start and end dates of menstrual periods and various associated symptoms. The app then uses machine learning algorithms to track periods accurately to identify patterns of changes within a cycle and predict the time of ovulation for those who want to conceive or avoid pregnancy. It is a period tracking app that uses AI to predict ovulation. The app also has a feature where like-minded individuals can chat and ask personal questions and a conversational Chatbot that helps the women understand their symptoms better.
2. *Clue* (Clue 2021): Clue is another popular app that uses machine learning on the collected data to study and monitor menstrual health. Clue also has plans to use algorithms to identify a women's risk of polycystic ovarian syndrome (PCOS) by analyzing irregular menstrual cycles and associated symptoms captured on the app.
3. *Ovia Health* (Ovia Health 2021): Ovia Health provides an app for reproductive health support enabling women to track fertility and manage pregnancy and newborn care. They have accumulated one of the world's most extensive reproductive and family health datasets and are employing predictive analytics to add value for their users.
4. *Glow* (Glow 2021): Glow uses the power of data science to empower users with information related to fertility (using the Glow app), pregnancy (using the Nurture app), newborn care (by using the Baby app and providing personalized information regarding baby's health), and sexual and menstrual health (using the Eve app).
5. *Natural Cycles* (Natural Cycles 2021): Natural Cycles is the first FDA-approved non-hormonal birth control app that uses temperature to enable women to track their cycle and predict fertile days.
6. *Ava* (Ava 2021): Ava is the first and currently the only FDA-approved wearable bracelet to predict the start and end of the fertile window. The Ava bracelet captures and analyzes skin temperature, resting pulse rate, heart rate variability ratio, blood perfusion rate, and breathing rate using AI algorithms. The product is backed by validation in clinical trials and several publications (Goodale *et al* 2019, Shilaih *et al* 2017, Shilaih *et al* 2018).
7. *Mira* (Mira 2021): Mira provides an FDA-approved personalized at-home hormone tracking system that consists of a urine test wand, Mira Analyzer, and the Mira app. The user inserts the wand into the analyzer, and the concentrations of the fertility hormones, namely estrogen and LH, are

detected. The information is then transferred to the app. The AI algorithms embedded in the app analyze the hormone levels to predict the fertility window with over 99% accuracy.

8. *Eveline Care* (Eveline 2021): Eveline Care provides an Eveline Ovulation test with 5 to 19 strips to track the LH cycle using urine samples, a smart scanner to help read the strip, and an app with an embedded AI algorithm to predict the fertile window. As with other AI-powered apps, the algorithms become more accurate over multiple cycles of use.

In general, these menstrual cycle tracking apps have a 5% typical-use failure rate and 1% perfect-use failure rate, making these techniques comparable to existing family planning methods such as the pill, vaginal ring, and other FABMs (Freis *et al* 2018). Moreover, these FemTech applications have made fertility management more affordable. With free versions and premium versions available for a few hundred dollars, these apps are relatively more affordable for women to try before they opt for mainstream infertility treatments that cost above tens of thousands of dollars. Further, the apps do not have the side effects associated with the use of conventional contraceptive pills and intrauterine devices.

Univfy (2021) provides a highly scalable AI platform that generates scientifically validated personalized reports to people of all demographics about their probabilities of getting pregnant with *in vitro* fertilization (IVF). The chances are based on age, body mass index (BMI), ovarian reserve results, semen analysis results, and reproductive history. The predictions of the algorithms are validated against the IVF outcomes of the treatment center.

With more women planning to have children at older ages, there is a demand for solutions that can predict a woman's ovarian reserve, which is a measure of the number and quality of ovarian follicles. Ovarian reserve is generally estimated by measuring several hormones. Anti-Müllerian hormone (AMH) is the most commonly tested hormone (Broer *et al* 2014), and the others are estradiol (E2), follicle-stimulating hormone (FSH), LH, prolactin (PRL), testosterone, thyroid-stimulating hormone (TSH), and free thyroxine levels. In the US, proactive fertility hormone tests are generally not covered by insurance, and these hormone tests could cost more than USD 1000 at the doctor's office. Recently, a study has shown that the levels of these hormones measured using at-clinic lab-based venipuncture and home-based finger stick sampling were highly correlated (Burke *et al* 2019). Modern Fertility now provides an option for at-home fertility hormone tests (Modern Fertility 2021). The user can do a simple finger prick and mail the blood sample to a lab. A physician will review the results, and the user's fertility profile and timelines are sent back to the user. They also provide an app for ovulation and cycle tracking and prediction of the most fertile days in a cycle.

Solutions have also been developed to understand male fertility. Mojo AISA is an AI-powered semen assessment technology that could be used in laboratories to determine the motility and morphology of sperm cells. The device includes custom-built robotics and AI-trained algorithms.

3.3.3 Menopause

Many women do not receive sufficient information from their healthcare providers about menopause. FemTech solutions designed to manage menopause could help provide the much-needed information, alleviate stigma around menopause and its symptoms, and provide a comfortable platform for women to discuss and manage symptoms with their healthcare providers. In recent years, several approaches have been taken to help women detect and manage menopause.

- *Detection*: Currently, there is no highly accurate test to diagnose the beginning of menopause in aging women. Luo *et al* (2020) proposed a smart suit made of sensors to capture skin temperature and relative humidity. They also developed and tested the utility of an algorithm to detect the duration, frequency, and intensity of menopausal hot flashes. They concluded that sensor-based smart clothes could be helpful for women to monitor menopause symptoms effectively.
- *Symptom management*: The Embr Wave Bracelet (Embr 2021) is based on a concept called Thermal Wellness to manage hot flashes, stress, and sleep. The water-resistant bracelet cools or warms the inside of the wrist-based on the user's needs and changes how the user feels without affecting the core body temperature. Grace is another similar wearable bracelet that is currently under development (Grace 2021). Unlike the Embr bracelet, Grace has predictive algorithms to automatically track the onset of hot flashes and provide a cooling effect. It also comes with an app to help the user track hot flashes. Caria, by Chorus Health (Caria 2021), is a community-oriented menopause tracking app that provides evidence-based approaches such as CBT, hypnotherapy, nutritional therapy, and mindfulness to help women manage menopause. The app also uses deep analytics to track and manage symptoms. Other similar community-based online menopause management systems are Lisa Health (Lisa Health 2021) and Elektra Health (Elektra Health 2021). Elektra Health and Gennev (Gennev 2021) also provide telemedicine services for women's midlife health.

3.3.4 Pregnancy and maternal health

Several FemTech solutions that assist mothers from the time of conception through pregnancy, labor and delivery, and across newborn care are emerging. Some of them are detailed below.

3.3.4.1 Preterm birth

Preterm birth is a critical global health issue and is the leading cause of neonatal death (Blencowe *et al* 2012). Preterm infants are often missed because of the incorrect calculation of the gestational age (GA). Correct recognition of preterm birth can lead to timely clinical interventions to reduce stillbirth and an eventual reduction in mortality and life-long morbidity. Current clinical methods of GA estimation are primarily incorrect, with a 95% confidence interval of 18 to 36 days around the estimated GA (Fung *et al* 2020). Therefore, machine learning prediction models have been developed to improve the GA estimations. Fung *et al* (2020) used ultrasound-based fetal biometric data such as head circumference, abdominal circumference, and femur length to develop and validate a predictive algorithm for GA. They observed that the algorithm could detect GA between 20 and 30 weeks with a 95% confidence interval within three days which is significantly better than what traditional methods could estimate. They plan to incorporate the algorithm in a web portal or on an app. In low and middle-income countries, early prenatal ultrasound might not be available. For these situations, Rittenhouse *et al* (2019) have proposed a list of parameters such as last menstrual period, birth weight, maternal height, hypertension, twin delivery, and HIV serostatus to develop prediction models to identify preterm newborns with over 94% accuracy. The model is helpful to estimate GA at birth where early ultrasound is not available. Once the approach is validated extensively, the algorithm could be embedded into smartphone apps used in remote settings.

3.3.4.2 Pregnancy care

FemTech is making great strides in developing solutions for prenatal screening, pregnancy monitoring to manage conditions such as gestational diabetes and hypertension, and postnatal and maternal care to help mothers with breastfeeding and newborn care. Social media, mobile health (mHealth) apps, and wearables are becoming increasingly common in pregnancy care. Chan and Chen (2019) conducted a meta-analysis and observed that these interventions could improve the overall care of pregnant and postpartum women. These apps and wearables helped with pregnancy weight control (Chen *et al* 2018), gestational diabetes management (Miremberg *et al* 2018), and asthma control (Zairina *et al* 2016).

Globally, 5%–8% of pregnant women develop gestational hypertensive disorders such as essential hypertension, gestational hypertension, and pre-eclampsia (PE), PE being the most dangerous. If left untreated, PE can lead to maternal and fetal mortality (Mol *et al* 2016). Remote monitoring (RM) solutions could help address this issue by offering the possibility of home-based continuous care. Lanssenns *et al* (2018, 2020) have conducted large-scale studies called PREMOM1 and PREMOM2 (currently enrolling participants) to study the value of RM versus conventional care for women with gestational hypertensive disorders. They asked the women in the RM group to use simple devices such as a blood pressure monitor, activity trackers, and weight scale and record the data. The data were analyzed using predictive algorithms to determine if there was a reason for alarm.

More recently, there has been an influx of at-home remote pregnancy monitoring passive wearable devices that aim to help expecting mothers manage pregnancies in the comfort of their homes in between office visits. One such example is the INVU system by the Nuvo Group which consists of a wearable sensor belt with eight biopotential sensors to capture the electrocardiogram and six acoustic sensors to capture the phonocardiogram signals (Mhajna *et al* 2020). These signals are transmitted via Bluetooth to a mobile device to cloud-based servers, where digital signal processing algorithms are used to calculate the maternal and fetal heart rates. Wireless remote monitoring of these heart rates could improve prenatal care, maternal and fetal outcomes, and provide a better labor experience. The FDA has cleared the INVU platform (INVU 2021). The company is building upon the platform to introduce non-invasive uterine activity monitoring to capture a comprehensive set of fetal parameters. These parameters, when used in AI-based predictive models, could assist in proactive pregnancy management. HeraBEAT is another CE and FDA-cleared smart fetal heart rate tracker that uses hypersensitive ultrasound Doppler technology to detect and track fetal heart rate and pulse (HeraBEAT 2021). The device is portable and easy to use at home without the need for medical professionals. It uses advanced algorithms to cancel the noise caused by fetal movements, device movements, and abdominal sounds.

In light of the COVID-19 pandemic, there has been an increased need for solutions that can allow physicians and nurses to monitor patients (COVID-19 infected or not) to minimize the risk of exposure. Royal Philips has developed a solution that consists of the Avalon Fetal and Maternal Pod and Patch, a single-use 48-hour disposable patch to capture the fetal and maternal heart rate and uterine activity without the need for constant repositioning by nurses (Philips Avalon Solutions 2021). The system also has an Avalon CL transducer to allow for cableless monitoring to give expecting mothers the freedom to move during labor. An Intellispace Perinatal information management system is also included to automatically display and stream patient data electronic medical records.

ExtantFuture is developing the world's smallest fetal monitoring patch called Modoo (Modoo 2021). It uses multi-channel high-sensitivity sensors to perform passive monitoring and a fetal monitoring algorithm to monitor fetal heart rate and movement. Labor begins with uterine contractions that prepare the cervix for the baby's delivery. Bloomlife has developed a solution for monitoring contractions, both Braxton Hicks and true labor (Bloomlife 2021). It consists of a Bloomlife accelerometer sensor, a continuous-use patch, and an app with embedded algorithms to measure and time contractions accurately. Bloomlife is also building a remote prenatal care platform that uses technology, data science, and clinical expertise to improve the quality of care and personalize prenatal care.

In terms of solutions for maternity care management in electronic health records, Dorsata provides a platform that could be embedded in an existing EHR tailored for Ob-Gyn workflow (Dorsata 2021). The intuitive platform is powered by an advanced clinical rules engine based on each patient's pregnancy history, risk factors, gestational age, etc, to improve diagnosis, treatment, and documentation.

3.3.4.3 Maternal care

The weeks following birth are a critical period for both the mother and the newborn, and postpartum care is as essential as pregnancy care. InTouch Health (InTouch Health 2021) provides neonatal telehealth to allow birthing facilities to remotely connect with neonatal specialists in real-time, who can then use live video and audio to assess the infant and provide specialty care. Telehealth services such as Mahmee (Mahmee 2021), Babyscripts (Babyscripts 2021), and Maven Health (Maven Clinic 2021) provide HIPAA-secure maternal and newborn care platforms driven by data insights.

Post-labor, the stress of labor and delivery can affect the pelvic floor muscles, having an effect on daily activities and also can result in urinary incontinence (Bø 2004). Pelvic floor issues occur in almost 70% of mothers (Price *et al* 2010). Elvie has come up with an intravaginal biofeedback-based pelvic floor trainer like a small pebble-shaped pod that women can place like a tampon (Te Brummelstroete *et al* 2019, Elvie Pelvic Trainer 2021). The trainer is made of medical-grade silicone. The trainer talks to an app that displays pelvic floor movements in real-time in response to the strength of the contraction as the user squeezes the pelvic floor muscles. The patented algorithms measure force and motion and detect incorrect contractions to provide feedback to the user to improve the technique and build muscle strength. Renovia is another company that has developed an FDA-cleared digital therapeutic pelvic floor muscle trainer called Leva (Renovia 2021).

MyMilk Lab (MyMilk 2021) has developed an innovative breast milk sensing device to sense the electrochemical properties of a small human milk sample to understand the nutritional profile. The details are transferred to the MyMilk cloud-based algorithm to identify milk status and early trends of diminishing milk supply, detect breast inflammation, and provide personalized guidance on managing supply and connecting with lactation consultants. Elvie has developed an innovative, small, lightweight, wearable smart breast pump that can be worn inside the nursing bra (Elvie Pump 2021). The accompanying app can monitor milk volume in real-time, track pumping history, and also helps to control the pump operation remotely. The in-app predictive algorithms also switch from the stimulation mode to the expression mode when it detects milk flow and will pause when the bottle is full.

Postpartum depression (PPD) is a mood disorder that affects mothers within the first year of delivery. It is highly prevalent and affects about 20% to 40% of new mothers in low- and middle-income countries (Gebregziabher *et al* 2020). Since mobile phone use has become increasingly prevalent globally, several studies have explored the use of innovative mHealth applications to prevent, detect, and manage PPD. Dosani *et al* (2020) reviewed literature related to mHealth technology for PPD management in low and middle-income countries. They concluded that the literature is limited, and there there is immense opportunity to develop applications for PPD by good collaboration between healthcare providers and app developers. In another similar study, Hussain-Shamsy *et al* (2020) observed that several mHealth tools are available to deliver multiple strategies to manage PPD, including psychoeducation, cognitive behavioral therapy (CBT), symptom monitoring, encouraging exercise, etc. One such example is the Nourish app (Nourish 2021), which provides bite-sized calming moments to mothers using mindfulness, yoga, breathing, and positive psychology techniques, etc. In another interesting research study proposed by Poudyal *et al* (2019), the use of wearable digital sensors to identify PPD risks and personalize treatment is being explored. The study plans to capture the behavioral patterns of mothers by analyzing the data generated by wearable digital devices in the following four domains: (i) physical activity data collected using mobile phone accelerometer, (ii) geographic range and routine of mothers collected using mobile phone-based Global Positioning System (GPS), (iii) the time and routine with their infants using proximity data collected via

Bluetooth beacons, and (iv) the verbal interaction between mothers and infants collected using phone audio recordings.

3.4 FemTech applications for non-reproductive health

3.4.1 Cancers of the breast, ovary, and cervix

FemTech solutions are emerging in cancer risk assessment and the early detection of breast, ovarian, and cervical cancers. Cervical cancer, a malignant condition of the uterine cervix, is most commonly caused by the human papillomavirus (HPV). Current screening is done using a Pap smear and hrHPV testing (USPSTF 2021). More than 80% of cervical cancers occur in low- and middle-income countries where delayed diagnosis and lack of treatment lead to poor prognosis. AI-based solutions have been explored to create accurate, affordable, and accessible cervical screening applications to serve these countries. For example, an AI-based image processing system has been developed at the Lehigh University in Pennsylvania (Xu *et al* 2017). The screening system uses cervigrams as the input to trained-image-based classifiers to detect early-stage precancerous growths. A cervigram is a digital photograph of the cervix, a more affordable option than Pap smears and HPV tests. In another study by Holmström *et al* (2021), a cloud-based deep learning system was used to successfully screen cervical smears captured using simple digital microscopy. MobileODT is a FemTech company that has developed the EVA system, which consists of a mobile FDA-cleared digital colposcope for magnified cervical visualization, an app, and a portal for HIPAA-compliant online data, images, and video management (MobileODT 2021). In collaboration with the National Cancer Institute, the company has developed a deep learning AI algorithm called AVE (Automated Visual Evaluation) to assess the digital images of the cervix for early signs of cancer (Hu *et al* 2019). After sufficient clinical validation, the AVE algorithm will be integrated into the EVA system.

Biological cycles affect most physiological and biological aspects of humans, and alterations of the circadian rhythm have been linked to various health disorders (Masri and Sassone-Corsi 2018, Sulli *et al* 2019, Cortés-Hernández *et al* 2020). Different biological parameters such as activity, sleep cycles, temperature, etc, follow circadian rhythmicity (a 24-hour cycle). These parameters could be measured and analyzed to capture the alterations in the circadian rhythms, thereby predicting the early presence of cancer. Hoogland *et al* (2019) observed that women with gynecological tumors in the ovary, endometrium, or uterus had lowered rhythmicity in the circadian activity patterns than women with benign tumors and suggested that wearable activity sensors could be effectively used to identify gynecologic cancers earlier.

Breast cancer is a malignant proliferation of cells within the breast tissue. Cyrcadia Health (Cyrcadia 2021) is another disruptive FemTech company that has developed an AI-based wearable breast health monitoring system called the Cyrcadia Breast Monitor based on the concept of circadian rhythm alterations in skin surface temperature. The system includes two wearable patches for each side of the breast embedded with temperature sensors and a Bluetooth recording device to capture the sensed temperatures and transfer them to the Cyrcadia app. The app sends the data to the Cloud, where an AI-based predictive algorithm detects the presence or absence of any abnormal changes in the breast tissue (Vinitha Sree *et al* 2020).

All cells release volatile organic compounds (VOCs) that may have an odor. When cells become malignant, these VOCs change, and the measurement of these changes can indicate cancer. A research group has developed a breath sensor in Israel that uses molecularly modified gold nanoparticles that provided a strain-related response when exposed to breath. The sensor and related algorithms could discriminate breath from ovarian cancer patients with 82% accuracy, even in the presence of environmental VOCs (Kahn *et al* 2015). More research and validation are required to bring such unique early detection tools to market.

Celmatix is a preclinical stage biotech company focused on tapping into the power of precision medicine, advanced machine learning and AI, and natural language processing (NLP) techniques to develop the next generation of interventions and advancements in ovarian health (Celmatix 2021). They operate a discovery platform that has helped them collect the largest structured dataset on ovarian health and identify targets for therapy. They partner with leading reproductive centers, drug discovery alliances, and pharmaceuticals to potentially create new therapeutics to address ovarian health issues. More such initiatives are needed to address other women's health-related issues.

3.4.2 Chronic and autoimmune diseases

Women's health goes beyond reproductive health and cancers. Autoimmune diseases such as lupus, multiple sclerosis, thyroid diseases such as Grave's disease and Hashimoto's thyroiditis, rheumatoid arthritis, and psoriasis are more common in women than in men due to changes in hormones and sex chromosomes. About 80% of those with autoimmune diseases are women (Angum *et al* 2020). Other chronic diseases such as osteoporosis (Pouresmaeili *et al* 2018), diabetes (CDC 2021a), and hypertension (Ahmad and Oparil 2017) are also more prevalent in women. Heart disease is the number one cause of death among women in the United States (CDC 2021b) due to risk factors such as severe obesity that arises from hormone changes, menopause, and pregnancy (Yang *et al* 2013).

FemTech solutions are being developed to help women manage chronic diseases. These solutions range from smartphone apps for education and awareness, health coaching for symptom and medication management, and technology-enabled solutions for accurate screening and diagnosis of chronic diseases. Mobile biosensors embedded in wearables and smartphones allow for frequent measurement of physiological parameters. In a study by Davergne *et al* (2020), machine learning-based predictive algorithms were developed to predict the disease flares in inflammatory arthritis using physical activity-related parameters such as the number of steps. Such algorithms embedded in wearables could help patients obtain better disease control through early identification of disease flares.

In an interesting study conducted at Stanford University, CA, it was observed that continuous monitoring of parameters such as heart rate, skin temperature, oxygen saturation level, activity-related parameters such as sleep, number of steps, and calories, acceleration forces, weight, and radiation exposure could provide insights into the presence/onset of inflammation, insulin sensitivity, Lyme disease, cardiovascular disease, and metabolic syndrome (Li *et al* 2017). Women-specific cloud-based health management systems are being established to collect and analyze such continuous real-time data to understand health conditions and outcomes better. These systems are referred to as the learning health systems (LHSs). One such proposed method is called myAva (Satveit 2017), which provides a model for precision health. As these data inputs are collected over time, patients and providers can collaborate to develop personalized treatment plans. The results of such plans for different women can also be added over time to a shared database. This shared database could help other patients to devise their unique treatment plans. Thus, a continuous improvement of healthcare can happen both at the individual level and at the community and global level. myAva is first being used to help women manage PCOS. myAva's Precision Nutrition Program is designed to provide a personalized nutrition plan for each individual based on her molecular profile. The plans can be altered according to the changing biomarkers that are monitored over time. Eventually, machine learning and deep learning algorithms could be effectively used to analyze the volumes of data collected by such LHSs to improve disease detection, management, and treatments.

In line with technologies based on precision health, LifeStory Health (LifeStory 2021) is developing a predictive technology for the early detection of several diseases based on the analysis of biomarkers in menstrual blood (Yang *et al* 2012). Wearable technologies such as CardioGuard, a smart wearable electrocardiogram monitoring sensor system, have also been explored to detect cardiovascular issues (Kwon *et al* 2014). Companies such as Bloomer Tech (Bloomer Tech 2021) are developing smart fabrics that integrate advanced fabrics, sensors, and machine learning algorithms to convert bras into heart health monitoring wearables. Such female-specific wearables have the potential to provide personalized healthcare management and early detection tailored for women.

3.4.3 Lifestyle/wellness

FemTech solutions that cater to the wellbeing of the mind and body by taking a multidisciplinary approach to holistic health are also being developed. The unique physiological and hormonal changes that occur during menarche, pregnancy, motherhood, and menopause lead to a predisposition to mental health issues in women. Since women in most countries still hesitate to seek in-person help and therapy, several tele-mental health platforms have emerged in recent years. One example is Maven Health (Maven Clinic 2021). Rose Health has developed a HIPAA-compliant mental health app and monitoring platform powered by AI and NLP techniques (Rose 2021). These advanced algorithms also allow Rose to detect early signs of anxiety, depression, and trauma.

Bellabeat is a well-known wearable smart jewellery available in the market for women (Bellabeat 2021). The water-resistant wearable can be worn as a bracelet, clip, or necklace. It helps track activity, sleep, menstrual cycle, stress and helps achieve overall wellness and fitness goals. The Bellabeat companion app syncs with Apple Health Kit and Google Fit apps. Bellabeat has built an extensive database and algorithms that help understand the correlations between women's hormonal cycle and wellness. These insights are being used to develop personalized wellness coaching programs that are adjusted around cyclical hormonal changes. The key concept is that women's wellness can be significantly improved when the wellness programs are in sync with the individual woman's natural cycle.

Several studies have also evaluated the use of mobile apps (Coelhoso *et al* 2019, Zhang *et al* 2019) or a combination of mobile apps and wearable fitness trackers such as Fitbit (Grossman *et al* 2018) and Onitor® Track (Buchan and Morgan 2020), to help with weight loss, and improve physical activity and stress management in women. All these studies have observed positive effects of these technologies on women's wellness.

3.5 What is next?

Female health is complex and is governed by several physical, emotional, and socioeconomic factors. A couple of decades ago, disease diagnosis, treatments, and prevention strategies were researched and developed using male clinical study subjects. Technology has a critical role in enabling the medical field to catch up with the research specific to female healthcare. As detailed in this chapter, technologies such as artificial intelligence, the Internet of Medical Things, wearables, social media, and virtual digital platforms have a significant role in helping us understand the underlying variables that cause differences between health outcomes in women and men. These FemTech technologies are helping to uncover new

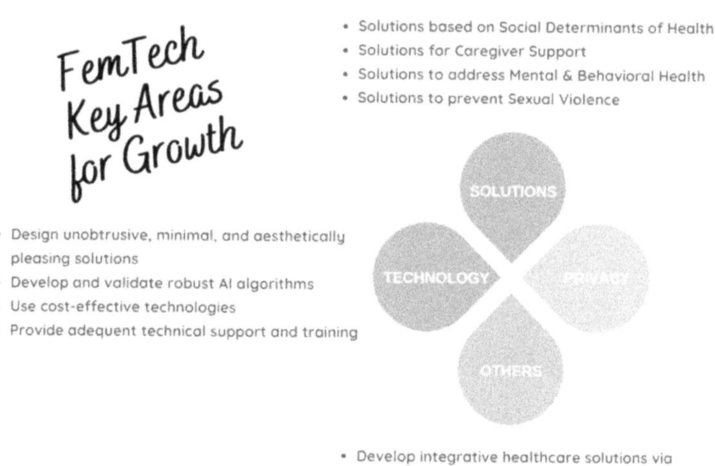

Figure 3.3. A few key areas for meaningful innovation in the FemTech space.

solutions for healthcare delivery, disease diagnostics, management, and treatment. Most of these digital FemTech platforms are powered by backend analytics that use technologies such as AI, natural language processing, and recommendation engines. Artificial intelligence and predictive analytics have become a mainstay in many FemTech solutions.

This chapter provided a review of FemTech solutions in reproductive medicine (menstrual health, fertility, menopause, pregnancy, and maternal care), non-reproductive women's health issues such as cancers of the breast, ovary, and cervix, chronic and autoimmune conditions, and general wellness. Several commercial applications in the market were mentioned, emphasizing how predictive analytics adds value to these applications. Below are a few of the general recommendations to improve applications and platforms in the FemTech space (figure 3.3).

3.5.1 Technology improvements

- *Validation*: The data are a critical component in the design of any digital health technology. The predictive algorithms used in any FemTech solution should be improved over time using the data collected during the subsequent use of the solution. New and emerging solutions with embedded AI have limited data on the accuracy of the predictions. These solutions should continue to be adequately validated in a timely manner through standard clinical trials and using the real-world evidence data collected. The clinical trials should be designed to address racial bias and other social determinants of health (SDOH). FemTech solutions can also help to recruit qualified patients for clinical trials because of their potential widespread use in all communities.
- *Design*: Wearables and smartphone-based solutions should have a design that is unobtrusive, minimal, and aesthetically pleasing to allow for comfortable wear in social settings. Solutions with mobile applications and web portals should have a user-friendly interface that is intuitive and captivating to maintain patient engagement.
- *Cost*: Currently, most FemTech solutions are sold to women directly. The cost of the solution should be affordable to users in all communities to provide value-based care.
- *Technical support*: Any FemTech solution should be integrated with adequate technical support, customer support, and clinical support. Both the patients and providers should have access to training to use the solution with high confidence.

3.5.2 SDOH-based care delivery

Examples of SDOH that affect healthcare delivery and outcomes include age, race, housing conditions, work, living conditions such as access to transportation, good food, childcare facilities, exercise, etc. FemTech solutions should leverage technology to incorporate SDOH in all phases of development to reduce racial health disparities and provide improved outcomes to all members of society equally.

Companies such as Wildflower Health (Wildflower Health 2021) have developed powerful digital engagement platforms to provide value-based care to mothers who have problems maintaining clinic appointments due to a lack of adequate transportation facilities and childcare. The platform leverages the data collected from remote monitoring and risk assessments, and applies analytics to predict user behavior and determine insights into the user's health.

The United States has the highest maternal mortality rate among 11 developed countries globally (TCF 2021), primarily due to the lack of adequate maternity care providers and midwives and the lack of comprehensive support during the postpartum period. The mortality rate in black women is at least three to four times higher than in other communities (CDC 2021c) because of the lack of culturally competent prenatal and postpartum healthcare. Culturally competent care often results in better patient engagement, favorable experiences, and improved clinical outcomes. Digital health solutions such as Candlelit (Candlelit 2021), Culture Care (Culture Care 2021), and Health in Her Hue (Health in Her Hue 2021) provide culturally competent healthcare access to Black women and women of color. Companies such as Folx Health (Folx Health 2021), Included Health (Included Health 2021), Queerly Health (Queerly Health 2021), Violet (2021), and Woven Bodies (2021) provide holistic healthcare support to the LGBTQ+ community.

Socio-cultural traditions in many parts of the world treat some women's health issues such as infertility, menstrual health, sexual wellness, and contraceptives as taboo subjects. Such norms prevent women from seeking timely and adequate care, leading to physical complications and mental health issues. Socio-cultural barriers to healthcare are highly prevalent in rural settings where women lack education and the mindset to be receptive to change to improve health outcomes. The limited or lack of availability of digital connectivity in these regions used to impact the efforts taken to serve these communities. However, the situation is improving in most parts of the world with higher use of smartphones and Internet connectivity. FemTech solutions should leverage this transforming digital landscape to provide adequate healthcare to women in rural areas via virtual telehealth services, remote monitoring, and AI-empowered clinical decision-support systems. Several such platforms have been mentioned earlier in this chapter. One such solution is CareMother (CareMother 2021) by CareNX Innovations. The company has developed Fetosense, a smartphone-based portable fetal heart rate monitoring kit that provides real-time alerts about abnormal findings using analytics. The company's AnandiMaa pregnancy care model for rural mothers uses a smartphone integrated antenatal care kit, a mobile app for monitoring and providing decision-support for high-risk pregnancies, and a web portal for comprehensive management of patient data.

Solutions should also be developed to bust myths and clarify taboos around female health issues and encourage women to participate in their self-care. FemTech solutions such as Poppy Seed Health (Poppy Seed Health 2021) could enable such patient engagement by providing 24/7 on-demand text access to doulas, midwives, and nurses to help women navigate pregnancy and postpartum health from the privacy of their homes.

3.5.3 Solutions for caregiver support

More than 60% of caregivers in the US are women (AARP 2021). They provide care to both children and aging parents. A few platforms such as Maven Health (Maven Clinic 2021) and Cleo (2021) are offering support to parents (working or not) by providing virtual personalized guidance on topics such as family planning, fertility, maternity, return-to-work and childcare, mental health support, etc. Low-cost voice-enabled interfaces such as Alexa, Google Home, and Siri use a combination of AI and NLP techniques to understand and sometimes even anticipate the needs of caregivers. The widespread adoption of telehealth and remote monitoring services that use AI to perform predictive and retrospective analytics is also a boon to caregivers. These technologies must be developed considering socio-cultural values (minority status, socioeconomic status) and should follow good privacy and security regulations (Lindeman *et al* 2020).

3.5.4 Solutions for mental health and sexual violence

Women are about two times more likely to suffer from depression than men (WHO 2021a). Although many FemTech virtual platforms provide mental and behavioral health support for women in all stages of life, only very few solutions have mental health as the core component. More solutions are needed to help women manage mental health issues arising from miscarriage, gender discrimination, eating disorders and poverty, and domestic and sexual violence. Real (2021) is an in-person and virtual therapy provider that empowers women with issues related to body image, intimacy, imposter syndrome, and concerns about having kids. Parla (2021) and Moment Health (2021) are a few other tele-mental health platforms.

Nearly one in five women experience some form of sexual violence during their lifetime (WHO 2013). Intimate partner violence (IPV) is one of the most common forms of sexual violence, and one in three women experience IPV. Callisto is a serial sexual assault prevention technology designed for serving survivors on college campuses (Callisto 2021). It is supported by a community of survivors, and is based on the concept that if one survivor reports unwanted sexual contact, others will speak up as well, and matching algorithms help detect the assaulter. myPlan is a free, private, secure, personalized, research-backed app developed by Johns Hopkins University, MD, to enable people to handle relationship abuse (myPlan 2021). It is an interactive safety decision aid to help the victim assess the risks and benefits and make informed decisions to protect their safety. While digital health solutions are emerging for screening and providing emotional health support for survivors of sexual violence and IPV, privacy and confidentiality should be critical components in designing such solutions to protect vulnerable women (Bacchus *et al* 2019, Emezue 2020).

3.5.5 Integrative women's health services

More work is needed in developing FemTech solutions that provide an integrative approach to healthcare. Such an approach is designed to discover the root cause of diseases and not just treat symptoms. It requires a multidisciplinary collaboration among a range of providers and the patient. For example, Tia provides a Whole Health Exam that includes annual primary care, gynecology exam, and mental health check-in to assess nine core health factors (reproductive health, genetic and family history, chronic stress, personal and social wellness, mental health, exercise, nutrition, sleep, and metabolism) (Tia 2021). The assessment helps to generate a personalized care plan for each woman. PA algorithms will play a critical role in studying patterns in such large diverse datasets. Such integrative digital health approaches should also be developed for the management of chronic diseases. For instance, for women living with osteoporosis, an app or telehealth platform that allows for effective communication between the patient, primary care physician, gynecologist, orthopedic doctors, and psychiatrists could prove beneficial in managing the disease.

Such integrated approaches should also be part of female-centric health apps for smartphones. For example, menstrual cycle tracking apps determine if the women have irregular menstrual cycle lengths. Although inconsistent use and reporting can lead to inaccurate cycle lengths, other potential health issues could be a more definitive reason for irregular menstrual cycles. Suppose these apps capture non-reproductive health-related symptoms such as blood sugar, blood pressure, etc. In that case, a deeper insight into underlying potential health issues can be obtained. Pattern recognition algorithms are well suited to determine correlations between all the parameters captured by these applications.

Science-backed alternative medicine approaches such as Ayurveda, yoga, and traditional Chinese medicine could also become part of these integrative approaches for providing preventative wellness services and treatments. For instance, Gynoveda uses technology and Ayurveda to provide solutions for menstrual issues (Gynoveda 2021). Tia also includes acupuncture sessions to help women manage headaches, stress, pre-menstrual symptoms, etc. Feminade (2021) is an online concierge service for women's hormone and reproductive health. It takes a holistic and root-cause personalized approach using an at-home hormone test kit and teleconsultations with licensed medical professionals, naturopaths, and nutritionists.

3.5.6 Regulation and privacy

This chapter outlined the potential of FemTech digital health solutions. However, digital solutions always come with issues surrounding data security and privacy. The Health Insurance Portability and Accountability Act (HIPAA) defines personal health information categorized as Protected Health Information (PHI). Most FemTech companies that collect and store personal health data fall outside the purview of HIPAA. HIPAA should be amended to minimize the potential data and privacy risks associated with FemTech solutions (Celia Rosas 2019). There should be better transparency of data use, and the users should have the ability to control how their data is used. Privacy and security should be considered while designing the technical architecture of any FemTech solution. If done correctly, it is possible to balance protecting users' personal information and promoting innovation.

References

AARP 2021 Caregiving in the US *Research Report* National Alliance for Caregiving https://aarp.org/content/dam/aarp/ppi/2020/05/full-report-caregiving-in-the-united-states.doi.10.26419-2Fppi.00103.001.pdf

Agarwal S K, Chapron C, Giudice L C, Laufer M R, Leyland N, Missmer S A, Singh S S and Taylor H S 2019 Clinical diagnosis of endometriosis: a call to action *Am. J. Obstet. Gynecol.* **220** 354.e1–354.e12

Ahmad A and Oparil S 2017 Hypertension in women: recent advances and lingering questions *Hypertension* **70** 19–26

Angum F, Khan T, Kaler J, Siddiqui L and Hussain A 2020 The prevalence of autoimmune disorders in women: a narrative review *Cureus* **12** e8094

Ava 2021 https://avawomen.com/

Babyscripts 2021 https://learn.babyscripts.com/our-solution

Bacchus L J *et al* 2019 Using digital technology for sexual and reproductive health: are programs adequately considering risk? *Glob. Health Sci. Pract.* **7** 507–14

Bellabeat 2021 https://bellabeat.com/

Blencowe H *et al* 2012 National, regional, and worldwide estimates of preterm birth rates in the year 2010 with time trends since 1990 for selected countries: a systematic analysis and implications *Lancet* **379** 2162–72

Bloomlife 2021 https://bloomlife.com/how-it-works/

Bloomer Tech 2021 https://bloomertech.com/#heartmonitor

Bø K 2004 Urinary incontinence, pelvic floor dysfunction, exercise and sport *Sports Med.* **34** 451–64

Broer S L, Broekmans F J, Laven J S and Fauser B C 2014 Anti-Müllerian hormone: ovarian reserve testing and its potential clinical implications *Hum. Reprod. Update.* **20** 688–701

Buchan K and Morgan H M 2020 Using the Onitor® Track for weight loss: a mixed methods study among overweight and obese women *Health Informatics J.* **26** 1841–65

Burke E E, Beqaj S, Douglas N C and Luo R 2019 Concordance of fingerstick and venipuncture sampling for fertility hormones *Obstet. Gynecol.* **133** 343–8

Callisto 2021 https://mycallisto.org/about

Candlelit 2021 https://livecandlelit.com/

CareMother 2021 https://caremother.in/

Caria 2021 https://hellocaria.com/

CDC 2021a Diabetes https://cdc.gov/diabetes/library/features/diabetes-and-women.html

CDC 2021b Heart disease https://cdc.gov/heartdisease/women.htm#how

CDC 2021c Maternal mortality https://cdc.gov/reproductivehealth/maternal-mortality/pregnancy-mortality-surveillance-system.htm

Celia Rosas 2019 The future is FemTech: privacy and data security issues surrounding FemTech applications *Hastings Bus. L. J.* **15** 319

Celmatix 2021 https://celmatix.com/

Chan K L and Chen M 2019 Effects of social media and mobile health apps on pregnancy care: meta-analysis *JMIR Mhealth Uhealth.* **7** e11836

Chen H, Chai Y, Dong L, Niu W and Zhang P 2018 Effectiveness and appropriateness of mhealth interventions for maternal and child health: systematic review *JMIR Mhealth Uhealth.* **6** e7

Cleo 2021 https://hicleo.com/

Clue 2021 https://helloclue.com/

Coelhoso C C, Tobo P R, Lacerda S S, Lima A H, Barrichello C R C, Amaro E Jr and Kozasa E H 2019 A new mental health mobile app for well-being and stress reduction in working women: randomized controlled trial *J. Med. Internet Res.* **21** e14269

Cortés-Hernández L E, Eslami-S Z, Dujon A M, Giraudeau M, Ujvari B, Thomas F and Alix-Panabières C 2020 Do malignant cells sleep at night? *Genome Biol.* **21** 276

Countrymeters 2021 https://countrymeters.info/en/World

Culture Care 2021 https://ourculturecare.com/

Cyrcadia 2021 http://cyrcadiahealth.com/ and https://cyrcadia.asia/

Davergne T, Rakotozafiarison A, Servy H and Gossec L 2020 Wearable activity trackers in the management of rheumatic diseases: where are we in 2020? *Sensors* **20** 4797

Dorsata 2021 https://dorsata.com/

Dosani A, Arora H and Mazmudar S 2020 mHealth and perinatal depression in low-and middle-income countries: a scoping review of the literature *Int. J. Environ Res. Public Health.* **17** 7679

DotLab 2021 https://dotlab.com/science

Earle S, Marston H R, Hadley R and Banks D 2020 Use of menstruation and fertility app trackers: a scoping review of the evidence *BMJ Sex. Reprod. Health* **47** 90–101

Elektra Health 2021 https://elektrahealth.com/

Elvie Pelvic Trainer 2021 https://elvie.com/en-us/shop/elvie-trainer

Elvie Pump 2021 https://elvie.com/en-us/about/company

Embr Lab 2021 https://embrlabs.com/pages/how-it-works-wave-2

Emezue C 2020 Digital or digitally delivered responses to domestic and intimate partner violence during COVID-19 *JMIR Public Health Surveill.* **6** e19831

Eveline 2021 https://evelinecare.com/

Feminade 2021 https://feminade.com/

Fitzpatrick M B and Thakor A S 2019 Advances in precision health and emerging diagnostics for women *J. Clin. Med.* **8** 1525

Flo Health 2021 https://flo.health/

Folx Health 2021 https://folxhealth.com/

Freis A *et al* 2018 Plausibility of menstrual cycle apps claiming to support conception *Front. Public Health.* **6** 98

Frost and Sullivan 2021 https://ww2.frost.com/frost-perspectives/FemTechtime-digital-revolution-womens-health-market/

Fung R *et al* 2020 Achieving accurate estimates of fetal gestational age and personalised predictions of fetal growth based on data from an international prospective cohort study: a population-based machine learning study *Lancet Digit. Health.* **2** e368–75

Gebregziabher N K, Netsereab T B, Fessaha Y G, Alaza F A, Ghebrehiwet N K and Sium A H 2020 Prevalence and associated factors of postpartum depression among postpartum mothers in central region, Eritrea: a health facility based survey *BMC Public Health* **20** 1614

Gennev 2021 https://gennev.com/

Glow 2021 https://glowing.com/apps

Goodale B M, Shilaih M, Falco L, Dammeier F, Hamvas G and Leeners B 2019 Wearable sensors reveal menses-driven changes in physiology and enable prediction of the fertile window: observational study *J. Med. Internet Res.* **21** e13404

Grace 2021 https://gracecooling.com/

Grandi G, Ferrari S, Xholli A, Cannoletta M, Palma F, Romani C, Volpe A and Cagnacci A 2012 Prevalence of menstrual pain in young women: what is dysmenorrhea? *J. Pain Res.* **5** 169–74

Grossman J A, Arigo D and Bachman J L 2018 Meaningful weight loss in obese postmenopausal women: a pilot study of high-intensity interval training and wearable technology *Menopause.* **25** 465–70

Gynoveda 2021 https://gynoveda.com/pages/about-us

Health in her hue 2021 https://healthinherhue.com/

HeraBEAT 2021 https://herabeat.com/

Holmström O *et al* 2021 Point-of-care digital cytology with artificial intelligence for cervical cancer screening in a resource-limited setting *JAMA Netw. Open.* **4** e211740

Hoogland A I, Bulls H W, Gonzalez B D, Wright A A, Kennedy B, Small B J, Chahal N, Arboleda B L and Jim H S L 2019 Differential patterns of circadian rhythmicity in women with malignant versus benign gynecologic tumors *Psychooncology.* **28** 643–46

Hu L *et al* 2019 An observational study of deep learning and automated evaluation of cervical images for cancer screening *J. Natl Cancer Inst.* **111** 923–32

Hussain-Shamsy N, Shah A, Vigod S N, Zaheer J and Seto E 2020 Mobile health for perinatal depression and anxiety: scoping review *J. Med. Internet Res.* **22** e17011

Included Health 2021 https://includedhealth.com/

InTouch Health 2021 https://intouchhealth.com/telehealth-solutions/

INVU 2021 https://nuvocares.com/solutions#

Jennings V, Haile L T, Simmons R G, Spieler J and Shattuck D 2019 Perfect- and typical-use effectiveness of the Dot fertility app over 13 cycles: results from a prospective contraceptive effectiveness trial *Eur. J. Contracept. Reprod. Health Care* **24** 148–53

Kahn N, Lavie O, Paz M, Segev Y and Haick H 2015 Dynamic nanoparticle-based flexible sensors: diagnosis of ovarian carcinoma from exhaled breath *Nano Lett.* **15** 7023–8

Kiira Health 2021 https://kiira.io/

Kwon S, Kim J, Kang S, Lee Y, Baek H and Park K 2014 CardioGuard: a brassiere-based reliable ECG monitoring sensor system for supporting daily smartphone healthcare applications *Telemed. J. E. Health.* **20** 1093–102

Lanssens D, Vonck S, Storms V, Thijs I M, Grieten L and Gyselaers W 2018 The impact of a remote monitoring program on the prenatal follow-up of women with gestational hypertensive disorders *Eur. J. Obstet. Gynecol. Reprod. Biol.* **223** 72–8

Lanssens D, Thijs I M and Gyselaers W 2020 Design of the Pregnancy Remote Monitoring II study (PREMOM II): a multicenter, randomized controlled trial of remote monitoring for gestational hypertensive disorders *BMC Pregnancy Childbirth* **20** 626

Li X *et al* 2017 Digital health: tracking physiomes and activity using wearable biosensors reveals useful health-related information *PLoS Biol.* **15** e2001402

LifeStory 2021 http://lifestoryhealth.com/

Lindeman D A, Kim K K, Gladstone C and Apesoa-Varano E C 2020 Technology and caregiving: emerging interventions and directions for research *Gerontologist.* **60** S41–49

Lisa Health 2021 https://lisahealth.com/

Luo J, Mao A and Zeng Z 2020 Sensor-based smart clothing for women's menopause transition monitoring *Sensors* **20** 1093

Mahmee 2021 https://mahmee.com/

Maijala A, Kinnunen H, Koskimäki H, Jämsä T and Kangas M 2019 Nocturnal finger skin temperature in menstrual cycle tracking: ambulatory pilot study using a wearable Oura ring *BMC Women's Health* **19** 150

Masri S and Sassone-Corsi P 2018 The emerging link between cancer, metabolism, and circadian rhythms *Nat. Med.* **24** 1795–803

Maven Clinic 2021 https://mavenclinic.com/

Mhajna M, Schwartz N, Levit-Rosen L, Warsof S, Lipschuetz M, Jakobs M, Rychik J, Sohn C and Yagel S 2020 Wireless, remote solution for home fetal and maternal heart rate monitoring *Am. J. Obstet. Gynecol. MFM.* **2** 100101

Mira 2021 https://miracare.com/how-mira-works/

Miremberg H, Ben-Ari T, Betzer T, Raphaeli H, Gasnier R, Barda G, Bar J and Weiner E 2018 The impact of a daily smartphone-based feedback system among women with gestational diabetes on compliance, glycemic control, satisfaction, and pregnancy outcome: a randomized controlled trial *Am. J. Obstet. Gynecol.* **218** 453.e1–53.e7

MobileODT 2021 https://mobileodt.com/

Modern Fertility 2021 https://modernfertility.com/science/

Modoo 2021 http://modoo-med.com/index.html

Moga M A, Bălan A, Dimienescu O G, Burtea V, Dragomir R M and Anastasiu C V 2019 Circulating miRNAs as biomarkers for endometriosis and endometriosis-related ovarian cancer—an overview *J. Clin. Med.* **8** 735

Mol B, Roberts C, Thangaratinam S, Magee L, De Groot C and Hofmeyr G 2016 Pre-eclampsia *Lancet* **387** 999–1011

Moment Health 2021 www.themomenthealth.com/

Moustafa S, Burn M, Mamillapalli R, Nematian S, Flores V and Taylor H S 2020 Accurate diagnosis of endometriosis using serum microRNAs *Am. J. Obstet. Gynecol.* **223** 557.e1–557.e11

MyMilk 2021 https://mymilklab.com/

myPlan 2021 https://www.myplanapp.org/en/our-story

Natural Cycles 2021 https://naturalcycles.com/

NextGen Jane 2021 https://nextgenjane.com/home

Nourish 2021 https://thenourishapp.com/theapp

Ovia Health 2021 https://oviahealth.com/

Parla 2021 https://myparla.com/

Philips Avalon Solutions 2021 https://www.usa.philips.com/healthcare/solutions/mother-and-child-care/fetal-maternal-monitoring

Poppy Seed Health 2021 https://poppyseedhealth.com/

Poudyal A, van Heerden A, Hagaman A, Maharjan S M, Byanjankar P, Subba P and Kohrt B A 2019 Wearable digital sensors to identify risks of postpartum depression and personalize psychological treatment for adolescent mothers: protocol for a mixed methods exploratory study in rural Nepal *JMIR Res. Protoc.* **8** e14734

Pouresmaeili F, Kamalidehghan B, Kamarehei M and Goh Y M 2018 A comprehensive overview on osteoporosis and its risk factors *Ther. Clin. Risk Manag.* **14** 2029–49

Price N, Dawood R and Jackson S R 2010 Pelvic floor exercise for urinary incontinence: a systematic literature review *Maturitas.* **67** 309–15

Queerly Health 2021 https://queerly.health/

Real 2021 https://join-real.com/

Renovia 2021 https://renoviainc.com/leva-digital-therapeutic/

Rittenhouse K J *et al* 2019 Improving preterm newborn identification in low-resource settings with machine learning *PLoS One.* **14** e0198919

Rock Health 2021 https://rockhealth.com/FemTech-is-expansive-its-time-to-start-treating-it-as-such/

Rose 2021 https://rosehealth.com/

Satveit S 2017 Addressing the unique healthcare needs of women: opportunity for change exists at the intersection of precision health and learning health systems *Learn Health Syst.* **2** e10033

Shilaih M, Goodale B M, Falco L, Kübler F, De Clerck V and Leeners B 2018 Modern fertility awareness methods: wrist wearables capture the changes in temperature associated with the menstrual cycle *Biosci. Rep.* **38** BSR20171279

Shilaih M, De Clerck V, Falco L, Kübler F and Leeners B 2017 Pulse rate measurement during sleep using wearable sensors, and its correlation with the menstrual cycle phases, a prospective observational study *Sci. Rep.* **7** 1294

Stephenson L A and Kolka M A 1999 Esophageal temperature threshold for sweating decreases before ovulation in premenopausal women *J. Appl. Physiol.* **86** 22–8

Sulli G, Lam M T Y and Panda S 2019 Interplay between circadian clock and cancer: new frontiers for cancer treatment *Trends Cancer.* **5** 475–94

Te Brummelstroete G H, Loohuis A M, Wessels N J, Westers H C, van Summeren J J G T and Blanker M H 2019 Scientific evidence for pelvic floor devices presented at conferences: an overview *Neurourol. Urodyn.* **38** 1958–965

TCF 2021 https://commonwealthfund.org/publications/issue-briefs/2020/nov/maternal-mortality-maternity-care-us-compared-10-countries

Tia 2021 https://asktia.com/

Univfy 2021 https://univfy.com/

USPSTF 2021 Cervical cancer: screening *Final Recommendation Statement* US Preventive Services Task Force https://uspreventiveservicestaskforce.org/Page/Document/Recommendation StatementFinal/cervical-cancer-screening2

Vinitha Sree S, Royea R, Buckman K J, Benardis M, Holmes J, Fletcher R L, Eyk N, Rajendra Acharya U and Ellenhorn J D I 2020 An introduction to the Cyrcadia Breast Monitor: a wearable breast health monitoring device *Comput. Methods Programs Biomed.* **197** 105758

Violet 2021 https://joinviolet.com/

Visana Health 2021 https://visanahealth.com/

WHO 2021a Depression https://who.int/teams/mental-health-and-substance-use/gender-and-women-s-mental-health

WHO 2021b Endometriosis https://who.int/news-room/fact-sheets/detail/endometriosis

WHO 2013 *Global and Regional Estimates of Violence Against Women: Prevalence and Health Effects of Intimate Partner Violence and Non-Partner Sexual Violence* (Geneva: WHO) https://who.int/reproductivehealth/publications/violence/9789241564625/en/

Wildflower Health 2021 https://wildflowerhealth.com/solutions

Woven Bodies 2021 https://wovenbodies.com

Xu T, Zhang H, Xin C, Kim E, Long L R, Xue Z, Antani S and Huang X 2017 Multi-feature based benchmark for cervical dysplasia classification evaluation *Pattern. Recognit.* **63** 468–75

Yang H, Zhou B, Prinz M and Siegel D 2012 Proteomic analysis of menstrual blood *Mol. Cell. Proteomics.* **11** 1024–35

Yang N, Ginsburg G S and Simmons L A 2013 Personalized medicine in women's obesity prevention and treatment: implications for research, policy and practice *Obes. Rev.* **14** 145–61

Zairina E, Abramson M J, McDonald C F, Li J, Dharmasiri T, Stewart K, Walker S P, Paul E and George J 2016 Telehealth to improve asthma control in pregnancy: a randomized controlled trial *Respirology* **21** 867–74

Zhang J, Jemmott and Iii J B 2019 Mobile app-based small-group physical activity intervention for young African American women: a pilot randomized controlled trial *Prev. Sci.* **20** 863–72

Chapter 4

Telemedicine

Vinithasree Subbhuraam and Debabrata Panigrahi

Telemedicine refers to the use of technology to deliver healthcare remotely. Recently, there has been explosive growth in the use of telemedicine, fueled by the COVID-19 pandemic. A telemedicine approach is appropriate for the prevention, diagnosis, treatment, and short-term and long-term management of several health conditions. The extensive use of the Internet of Medical Things (IoMT), wearables, and digital apps in the delivery of telehealth services has resulted in enormous data generation. The use of predictive analytics (PA) and artificial intelligence (AI) has therefore become an integral part of telemedicine to provide better and quicker diagnosis, assist in remote monitoring, ascertain the correct dose of medication, and, more importantly, predict critical medical events. This chapter provides a general idea of how telehealth services are enabled in radiology, cardiology, psychiatry, dermatology, ophthalmology, obstetrics, oncology, nephrology, rehabilitation, and pediatrics. Several commercial applications, in particular those that are AI-powered, are described. Each section also highlights how telehealth is perceived by providers and patients and the areas that need further research. Finally, a few general recommendations to improve telehealth adoption are presented.

4.1 Introduction

The concept of telemedicine evolved with the advent of faster and more effective telecommunication technologies and the pressing need to make healthcare accessible to all. With emerging tools and technologies, telemedicine has transformed into a complex integrated service used in healthcare facilities. It is a powerful technology that can be leveraged to facilitate remote healthcare services successfully by educating patients and healthcare professionals.

The last decade saw a significant development in PA, AI, and machine learning (ML) technologies, which substantially affect healthcare management. AI's ability to predict unprecedented events by leveraging human decision making has proven to be effective in healthcare. From replacing sophisticated medical instruments with

doi:10.1088/978-0-7503-2312-3ch4

relatively simple AI-enabled IoMT devices to the thorough analysis of electronic health records (EHR) using natural language processing (NLP) methods, AI is proving to be a significant boon to the healthcare industry. The latest report by Fortune Business Insights (FBI 2021) indicates that the global telehealth market was valued at USD 61.40 billion in 2019 and is projected to reach USD 559.52 billion by 2027. Because of the global COVID-19 pandemic, there has been a significant rise in the preference for e-health visits which will continue to enable the growth of several telemedicine innovations.

There has been an increase in the affordability and accessibility of essential telemedicine tools worldwide in recent years. People have access to advanced technology to facilitate live video telemedicine and are also acquainted with online video conferencing tools such as Skype, Microsoft Teams, Google Meet, etc, that are technically available 24/7. Telemedicine was initially adopted to provide primary healthcare to remote locations and address the shortage of healthcare professionals. It has now become an essential medium for delivering medical care more conveniently. With so many technological advancements and the rise in digital awareness, people have begun to prefer telemedicine over regular visits to get immediate and easy access to healthcare to address minor yet urgent complications.

This chapter provides a brief overview of telemedicine and PA, and how various healthcare departments have adopted telehealth services. The chapter will also review the advantages and challenges in implementing PA solutions for telemedicine.

4.2 Telemedicine and predictive analytics

Chronic disease has become the most complex and common global healthcare crisis. Management of chronic conditions requires a multidisciplinary approach to assess the patient's condition, preferably in real-time, and facilitate appropriate and timely interventions (Kuziemsky *et al* 2019). Protocols and continued care are an integral part of providing healthcare. Any deviation from this affects the quality of healthcare provided and endangers the patient's life, giving rise to unprecedented emergency conditions. Telemedicine could address this need for continuous monitoring by establishing a robust hassle-free communication methodology between the various elements of the health triaging system. However, sometimes not all aspects of a healthcare delivery system are available at all locations. In such scenarios, AI and IoMT solutions can provide remote care, intelligent predictive and diagnosis systems, and facilitate patient care through various virtual platforms.

The rise of digital health solutions has resulted in a surge in the volume of health datasets accumulated from patients and healthcare providers. Big data is being generated due to a significant increase in IoMT devices and apps and electronic medical and health records. This big data is characterized by (i) *volume*, (ii) *velocity* due to the rapidly increasing speed at which datasets are accumulating, (iii) *variety* due to the diversity of data types present in healthcare datasets, (iv) *variability* due to the dependence of the meaning of the data on the care setting, physicians, and patients, and (v) *value* due to the need to capture accurate data that is meaningful for a particular healthcare objective to derive useful information that supports the intent. Big data analytics are a group of techniques based on PA, AI, ML, deep learning, etc, which can help examine these large volumes of rapidly accumulating valuable and variable healthcare data in the proper context. These techniques could help uncover patterns, correlations, and trends to gain contextual insights. When combined with real-world patient data, these techniques could also enable the discovery of personalized treatment plans for individuals or patient cohorts.

A generic telemedicine architecture consists of the following: data are collected from the patient and securely transferred to a remote clinical back-end server in the Cloud via the Internet. Physicians could access this data, perform analysis with/without a set of AI-based computer-aided clinical support systems, and report the results back to the patient or the patient's local physician for further steps. The following section provides a detailed overview of the different types of telemedicine solutions.

4.3 Types of telemedicine solutions

4.3.1 Live or synchronous telemedicine

Synchronous telemedicine is a real-time telehealth solution with a two-way communication between patients and medical practitioners, generally via video-based communication. Special telehealth-enabled instruments, such as a video otoscope or an electronic stethoscope, could also be used for a virtual examination. This telemedicine solution is well suited for busy patients and proves to be a good substitute for in-person doctor visits. To make this technology feasible and accessible to everyone, telehealth providers have developed several technology-based and device-based solutions. A large amount of data is generated in the process and archived as electronic health records, analyzed by AI-based algorithms to generate reports and help physicians diagnose effectively.

For example, Medweb uses radiology information system (RIS)/picture archiving and communication system (PACS) enabled technology to collect patient data to be able to provide services even in the most remote locations such as Afghanistan (Medweb 2021). Lemonaid Health uses machine learning algorithms and physician recommendations to offer quick and inexpensive telehealthcare and prescription delivery (Lemonaid Health 2021). eVisit provides a virtual health platform that enhances clinical workflows such as virtual visits, scheduling, intake, waiting room management, discharge, and data analytics, keeping patient data secure and connecting patients to healthcare providers (eVisit 2021).

Many healthcare providers and patients embrace this type of telemedicine service due to its ease of use and reasonable costs. Service providers also strive to be Health Insurance Portability and Accountability Act (HIPAA) compliant to protect patient data privacy.

4.3.2 Asynchronous telemedicine

Asynchronous telehealth refers to the 'store and forward video conferencing' means of providing telehealth services. In this type of service, medical data are acquired and transmitted to the medical practitioner at a convenient time for offline analysis. The service provider ensures that adequate patient images, videos, or health information are captured and sent to the medical practitioner for a diagnosis. This method eliminates the need for sophisticated technologies for continuous data transmission and is used widely in remote locations with a scarcity of specialized physicians. For example, the nonprofit organization ORBIS links clinicians in developing countries with mentors in developed countries to improve the diagnosis and management of ocular diseases (Orbis 2021). In recent years, this telehealth service has been used commonly to treat ocular disorders and skin disorders and diagnose diseases in radiology data such as x-rays, CT scans, etc.

For example, in radiology, medical professionals at remote locations can acquire and transmit radiological images to specialists at other sites via the asynchronous set-up. In dermatology, a similar arrangement could be used to send patients' skin images to specialists. In ophthalmology, retinal images of patients suffering from diabetic retinopathy could be transferred to specialists for further diagnosis and treatment recommendations.

With the advent of AI-based prediction models, these telehealth services also provide options to help patients identify their potential ailment and probable causes. For example, Healthtap (Healthtap 2021) has developed a chatbot-based mobile app called HealthTap AI that contains an in-built recommendation engine built using AI. When a user describes their symptoms in the app, the app then generates a list of potential causes for the symptoms and recommends the next steps for taking care of them. Ada (2021) is another similar symptom checker app developed using physician input and pattern recognition algorithms. The asynchronous telehealth delivery model provides personalized healthcare using limited resources and cognitive user interfaces.

4.3.3 Remote patient monitoring

Remote patient monitoring is a telemonitoring method used by healthcare professionals to remotely track a patient's vital signs and symptoms. This type of solution is beneficial for conveniently monitoring patients with chronic conditions such as heart disease, diabetes, chronic obstructive pulmonary disease, etc, and the elderly with neurocognitive disorders such as Parkinson's and Alzheimer's disease.

Patient monitoring is one of the first and most common applications of telemedicine. Remote monitoring allows for a faster and cost-efficient way of conducting regular doctor-to-patient consults to assess the patient's current state and clinical results. Data from the digital sensors that capture the patient's vital signs can be analyzed by an app using AI-based solutions. Any significant change in signs can automatically alert the healthcare providers, who can then consult with the patient via face-to-face video conferencing to alter medications or treatments. Emergency alerts can also trigger ambulatory services to attend to the patient in distress. This model's objective is to provide accessibility, ease, efficiency, and reduced costs compared to physical patient monitoring.

The data resulting from continuous remote monitoring of patients with chronic conditions could further improve predictive algorithms. Such datasets could provide valuable insights into how vital signs change over time and even how various vital signs correlate. Another benefit is to provide better remote wound care. Software apps installed on a patient's phone can analyze wound images and calculate the rate of change of wound size and appearance and send the information to the physicians. The physicians can then alter medications and treatment remotely.

Advances in AI and robotics have also led to the development of telepresence robots that can examine the vital signs of patients remotely. Such technologies are very beneficial in a pandemic situation where these robots can reduce healthcare staff's exposure to infectious diseases. The robots are designed to be lightweight, foldable, and stable and can move quickly between different patient rooms. They are embedded with audio and video devices of superior quality. They are also designed to be HIPAA compliant to protect the patient's privacy and data security. These robots can be used to monitor and treat patients remotely. One example is the Ohmni Telepresence Robot (Ohmni 2021). The robot can autonomously move around hallways and rooms through a remote-controlled software interface that uses a combination of artificial intelligence algorithms and vision systems for navigation and obstacle detection. The robot has a screen for patient–doctor communication. These robots have been a boon to healthcare providers during the COVID-19 pandemic, not only for diagnosis but also for screening, disinfection, surgery, telehealth, logistics, and social care (Shen *et al* 2021).

4.3.4 Mobile health (mHealth)

The increasing use of mobile devices and their high user acceptability rate have paved the way for mHealth. This form of healthcare typically involves gadgets such as mobile devices, tablets, personal digital assistants, and various other wireless devices to provide an easy and efficient user interface facilitated by satellite communication to transfer information effectively. The gadgets can have health-based apps installed to monitor everything ranging from daily steps and water intake to heart rate and blood sugar levels. These apps can encourage healthy living and be linked to the patient's healthcare provider's records. The provider can use this data as a type of patient history and use it for remote monitoring.

Another significant benefit of such health apps is promoting information about a public health crisis such as the COVID-19 pandemic and keeping the population updated. mHealth technologies should incorporate live technical support, be easy for users, and include face-to-face communications. If the apps are affordable, they will prove very valuable in improving healthcare access in rural settings.

mHealth has been instrumental in providing patients and their healthcare providers easy access to information through the web or mobile-based apps. The patients can book appointments easily, have their prescribed drugs delivered, and, most importantly, establish a secure communication pathway with the service providers to ensure hassle-free services. In some apps, the patients can connect to other patients with similar health conditions through social networks, enabling a health support system.

4.4 Applications

4.4.1 Teleradiology

Radiology is a field of medicine that uses medical imaging to diagnose, manage, and treat several types of diseases. With the advancements in communication systems, encrypted information transfer techniques, and image processing techniques, physicians and hospitals are embracing the concept of teleradiology (Bashshur *et al* 2016). Teleradiology refers to the idea of transferring medical images from a primary system to a remote location, where the radiologist receives and interprets images. Teleradiology has evolved in a very short period with the use of AI tools and algorithms.

4.4.1.1 General framework

The main components of a teleradiology system are (i) an image acquisition station, (ii) an image transmission network, (iii) an image receiving station, and (iv) image analysis software to augment the radiologist's interpretation. Figure 4.1 shows an example of a typical teleradiology framework, and the sections below describe each of the components in more detail.

(i) *Image acquisition station.* Teleradiology can be used to interpret images from digitized x-rays, ultrasound scans, computer tomography (CT), magnetic resonance imaging (MRI), nuclear imaging, etc. An internationally accepted standard known as DICOM, which stands for Digital Imaging and Communications in Medicine, is used to handle, store, transmit, and print medical images and metadata. DICOM regulates the standard for file format and communication protocols. DICOM's two critical definitions are grayscale softcopy presentation state (GSPS) and grayscale standard display function (GSDF). The GSPS is an object for

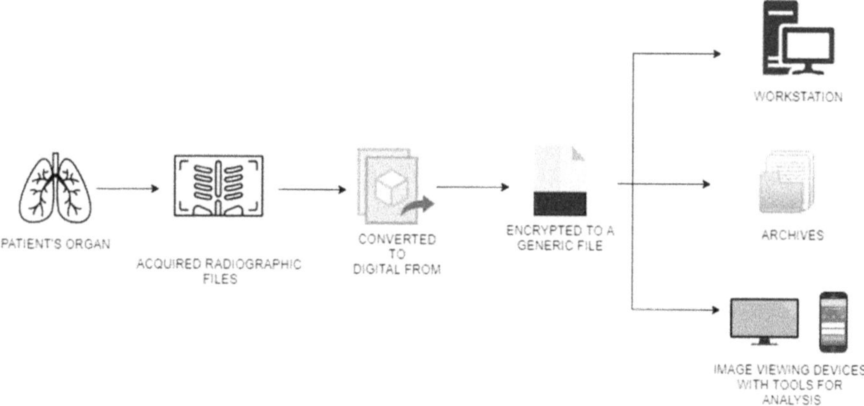

Figure 4.1. A secure teleradiology framework for transmitting DICOM images.

storing and communicating the parameters that describe how an image should be displayed on a softcopy. Parameters include grayscale transformations, shutter transformation, image annotation, spatial transformations, etc. Unlike GSPS, GSDF can be seen as a method that improves information perception and enables consistent image presentation across multiple displays. GSDF mitigates the problem of various displays from several service providers, each with varying luminance ranges and white points.

(ii) *Image transmission network.* PACS is a modality of imaging that helps transmit images from the acquisition site to multiple remote locations for analysis. It is usually an integration of image acquisition and storage devices, transmission networks, display stations, cameras, and techniques to integrate with the hospital's RIS and hospital information system (HIS). RIS is a crucial component of the imaging enterprise that manages imaging workflow, images, data, billing information, and general record keeping. PACS, image compression technologies, wireless transmission, and portable viewers have helped teleradiology to become more accessible and feasible to physicians. To protect confidential medical records from being accessible to unauthorized individuals, teleradiology service providers also follow security measures such as encryptions and establishing a virtual private network (VPN) between hospitals and receiving locations.

(iii) *Image receiving station.* Image receiving stations are equipped with (i) a modem, (ii) computer hardware, (iii) image-enhancement software, and (iv) monitor(s). The modem's maximum speed affects the time of data transmission from the transmission station. The receiving station must also be equipped with a modem of maximum speed equal to or greater than that of the transmitting station. Data are stored either in a compressed or uncompressed format on hard disks. A majority of teleradiology systems have grayscale window/level and magnification image-enhancement software. This software can also perform grayscale mapping, positive–negative reversal, annotation, minification, edge enhancement, image flip/rotate, and histogram equalization. In terms of monitors, the most commonly used monitors are of the resolutions 512*512 to 2048*2048, and screen sizes range from 14 to 21″. Monitors with high foot-lambert values are preferred for high differential contrast between the shades, making it easier for human eyes to read images.

(iv) *Image analysis software.* Image analysis software is critical for enabling the optimal display of digital images on electronic displays and improving transmission, storage, and retrieval speeds, all of which are essential to teleradiology. Analysis software also has image processing algorithms based on AI and predictive analytics to help radiologists achieve a more objective evaluation. For example, Georgiadis *et al* (2007) designed and implemented a personal digital assistant (PDA)-based teleradiology system with a DICOM-server connected to an MRI unit and three PDAs. Image analysis software was installed in the PDAs. The software had a processing component to manipulate grayscale images and filter using spatial filtering techniques. It also had an image analysis component that could classify three major types of brain tumors using a probabilistic neural network (PNN) classifier and image texture-based features. Recent advancements in 3D imaging have paved new ways for adaptive and context-based compression strategies to allow adequate storage and transfer of segmented binary data (Aldemir *et al* 2020).

4.4.1.2 Perceptions on teleradiology practice

The use of advanced instrumentation and complex systems in radiology has made its implementation a challenge. Therefore, adequate medical imaging-based procedures are inaccessible to nearly two-thirds of the world's population (WHO 2012). Teleradiology can bridge this gap. There have been multiple clinical trials and user satisfaction studies related to teleradiology. Rosenkrantz *et al* (2019) conducted a recent review to explore the current state of teleradiology practice and the perceptions of radiologists. They concluded that despite various challenges, teleradiology has now been integrated with conventional radiology. Teleradiology has enabled radiologists to provide and receive multispeciality services irrespective of geographical location and time with reduced turnaround times. But they also pointed out the need for quality assurance of off-site examinations. Based on surveys conducted with 87 teleradiologists and 731 non-teleradiologists, the most important conclusion of the study is that the advancements benefited radiologists in smaller practices in teleradiology less than their counterparts in primary facilities. This study also indicated that using information technology integration services at appropriate places could mitigate this challenge.

4.4.1.3 Commercial applications

1. *Nines* (2021): Nines is an AI-powered radiology-as-a-service teleradiology solution. NinesAI, an FDA-cleared software feature of Nines, facilitates the automated review of head images for the presence of intracranial hemorrhage and masses. The software can help radiologists prioritize the cases based on severity. For example, the software can alert the radiologists about an image with a potentially life-threatening brain bleed so that the patient can be reviewed immediately. Radiologists can thus deal efficiently with an increasing case-load in situations such as COVID-19, attending to high-risk cases first.
2. *Paxera Health* (2021): Paxera Health, a leading medical imaging solution developer based in Boston, MA, offers a cloud-based teleradiology solution. The users can send studies to multiple radiologists in seconds and receive interpretations within minutes.
3. *MILLENSYS* (2021): The MILLENSYS platform is another web-based and Cloud-based teleradiology framework that can provide remote diagnosis and reporting using their Vision Tools Workspace solution. For example, they have an FDA-cleared Vision Tools MammoView mammography PACS

workstation for handling mammography images. Combining this solution with their cloud-based teleradiology framework allows for faster analysis of images.

4. *DeepTek* (2021): DeepTek is another AI-powered teleradiology solution provider that leverages AI to provide 24/7 image analysis and reporting services. They use innovative imaging algorithms powered by deep learning to analyze radiographs, CT scans, and MRI.

4.4.2 Telecardiology

Cardiology, among all other clinical disciplines, has been the epicenter of telemedicine. Due to the complex nature of cardiological conditions, there is a need for consultation with specialists in the field to either obtain a second opinion or provide care to people in remote locations.

4.4.2.1 General framework

In general, a telecardiology framework involves using a 12-lead electrocardiogram (ECG) in the transmitting center, such as a primary care provider's office or a rural clinic. The acquired ECG signals are transferred to the receiving end, which is usually the specialist's office. Once the specialists interpret the signal, an oral or written report is sometimes sent back to the transmitting center. The ECGs are also stored as part of the electronic medical record for the patient for future reference. In addition to ECG machines, which could be standard ECG or wearable ECGs, the transmitting centers could also be equipped with an echocardiogram. This ultrasound machine can capture images of the structure and function of the heart.

There are three aspects to a telecardiology system (Molinari *et al* 2018). Pre-hospital telecardiology helps remote monitoring of patients with acute coronary syndrome and for use by primary care physicians and general practitioners. In-hospital telecardiology primarily deals with real-time transfer and interpretation of electrocardiography signals between rural and tertiary care hospitals. Post-hospital telecardiology comprises solutions for monitoring discharged patients and reducing re-admissions in patients with heart failure or implants.

4.4.2.1.1 Pre-hospital telecardiology

Pre-hospital telecardiology services are the services provided to monitor critical cardiac conditions such as myocardial infarction (MI) which have a better prognosis if detected and treated early. In the case of MI, the time taken from the symptom onset to the performance of perfusion defines the infarct size, length of hospitalization stay, and mortality rate. The American Heart Association defines this interval as no more than 90 min (ACC 2004), but most centers are not equipped to treat patients within this time frame. In recent years, it has been determined that there is a good agreement between tele-ECG and standard ECG regarding picking up changes (Schwaab *et al* 2005). Therefore, telecardiology services that can capture and transfer ECG to a specialist for quick interpretation can help faster triage and faster interventions. For example, in a recent study Saberian *et al* (2019) assessed and concluded that pre-hospital triage via telecardiology could reduce the coronary reperfusion time in MI patients. This service is specifically beneficial for patients living in rural areas who do not have access to specialist cardiac care within a short time. Recently, De Bonis *et al* (2021) published their experience using telecardiology during the COVID-19 pandemic in Italy. Their established telecardiology system is composed of a network of hospitals and ambulances to allow for the rapid treatment of patients with MI. Despite the pandemic, they observed that their telecardiology system continued to perform efficiently. Telecardiology can also help general practitioners diagnose chest pain accurately, reduce unnecessary hospitalizations, enable quick interventions, and reduce healthcare costs (Molinari *et al* 2004).

4.4.2.1.2 In-hospital telecardiology

The majority of in-hospital telecardiology involves real-time echocardiography transfers between primary and tertiary care specialty centers using synchronous or asynchronous transmission modes. Tele-echocardiography is capable of evaluating a variety of cardiac and vascular structures and physiological functions. Because of the advancements in real-time ultrasound image transmissions, professional cardiologists can remotely assist a sonographer in conducting the examination and proposing diagnostic and therapeutic solutions. Tele-echocardiography has been valuable in diagnosing and managing congenital heart disease in newborns (Krishnan *et al* 2014) and managing heart failure in adult patients (Hjorth-Hansen *et al* 2020). Hsieh and Lo (2010) suggested a model where 12-lead ECGs could be integrated into a unified format based on DICOM and stored in the PACS. In an emergency teleconsultation, remote specialists can refer to patients' ECG, chest x-ray, and other relevant images from a single source and make a diagnosis.

4.4.2.1.3 Post-hospital telecardiology

Post-hospital telecardiology is generally provided for telemedicine-based rehabilitation after cardiac surgery. Compared with conventional approaches, telecardiology was found to be more suitable for patient care in cardiac rehabilitation for risk-factor modification and exercise monitoring for patients (Scalvini *et al* 2009). Telecardiology-based follow-up of patients discharged after acute coronary syndrome has shown reduced hospital re-admissions (Chiantera *et al* 2005). Further, tele-support of patients after an acute episode of MI significantly improved their survival rates at one year (Roth *et al* 2009).

Due to the aging population and the fact that heart failure (HF) is the leading cause of hospitalization among patients above 65, HF can contribute to an increased healthcare burden in terms of hospitalization costs and mortality rate (Jackson *et al* 2018). Therefore, there is a need for solutions to improve HF prevention, management, and monitoring. A recent clinical trial of 3230 patients with diabetes, chronic obstructive pulmonary disease, or HF demonstrated that telehealth interventions helped patients better manage their conditions at home. Telemedicine also enables healthcare providers to make better-informed decisions on where to refer or admit the patient, and therefore reduce the need for unnecessary emergency admissions overall (Steventon *et al* 2012).

With the increased usage of implantable devices such as pacemakers, cardiac resynchronization therapy, and implantable cardioverter defibrillators, there has been a significant increase in the number of follow-up visits to check the proper functioning of these devices. Telecardiology evaluations could help avoid these time-consuming and costly visits. In the COMPAS trial, Mabo *et al* (2012) studied the safety of the long-term remote monitoring of pacemakers and concluded that telecardiology reduced the number of unwanted ambulatory visits. De Ruvo *et al* (2011) compared the remote monitoring and in-office visit experiences of 99 patients with implantable defibrillators for cardiac resynchronization therapy. They observed that the patients who had only quarterly in-office visits without remote monitoring had a significant 86% higher risk of delayed detection of adverse reactions of using the implantable device.

4.4.2.2 Perceptions on telecardiology practice

Telecardiology empowers primary care practitioners to diagnose and treat cardiac patients accurately. In the process, telecardiology also helps increase the clinical knowledge of the practitioners. Since a single-lead, wearable ECG device is probably more efficient than a Holter machine in detecting and tracking arrhythmias continuously, telecardiology set-ups can significantly reduce missed cardiac events. Further, for patients with non-life-threatening cardiac situations, a telemedicine-based intervention could save them thousands of dollars they could have otherwise spent on an unnecessary visit to an emergency room. Telemedicine could also help patients stay up-to-date on prescriptions and track their cardiac health more effectively.

Further studies are needed to assess how practitioners can effectively embrace telecardiology at primary healthcare centers. Successful integration would require training on computers and telemedicine laws and practices. As discussed in the next section, several AI-integrated telecardiology solutions are being developed and offered to healthcare providers and patients. These solutions need adequate validation with large-scale clinical studies to test the hardware and network architecture and validate the robustness of the predictive algorithms.

4.4.2.3 Commercial applications

1. *Coala*: Coala Life has launched Coala Connect Rx, a telemedicine-based prescription in partnership with leading cardiologists. Users receive a Coala Heart Monitor, a CE and FDA-cleared patch-free, smartphone-based ECG monitor that monitors and detects the most common cardiac arrhythmias in real-time. The data are sent to both the patients and the healthcare providers. The data are also analyzed by predictive algorithms based on pattern recognition, P-waves, and RR variability in a Cloud-based platform (Insulander *et al* 2020). Atrial fibrillation (AF) in stroke survivors is typically evaluated by short-term ECG monitoring in the stroke unit. Long-term continuous ECG monitoring using chest and thumb ECG could enable quick AF detection and prompt anticoagulation treatment to prevent recurrent stroke. In a recent study, called the TEASE study (Magnusson *et al* 2020),

the use of the Coala Heart Monitor by 100 stroke patients was evaluated. AF was detected in 9% of the patients using the monitor twice a day for a month. The team recommended that the monitor could be a feasible evaluation tool for stroke survivors.

2. *Reid Telehealth Service* (Reid 2021): Reid Telehealth Service allows patients with MI or HF to receive telecardiology services from their board-certified cardiologists. This service is beneficial for patients in rural communities. The cardiologists evaluate the patients remotely using a digital stethoscope, exam camera, and ECG. An onsite telepresenter who is a trained healthcare professional remains with the patient to assist with the examination.
3. *InTouch Health* (InTouch 2021): The InTouch Cardiology solution supports even the most complex cardiac patients because of the use of peripheral-enabled devices. The solution is built on the Solo™ Cloud-based platform, which provides an integrated and scalable care experience. The platform has several modules for coordinating patient intake, device scheduling, clinical documentation, and handling patient's digital images. The platform also has in-built business analytics programs to drive operational and financial improvements.
4. *Implicity* (Implicity 2021): Implicity is a French digital medical technology company that provides an innovative global remote monitoring platform for connected medical devices. The platform includes AI-based algorithms that use the collected data to enhance patient care. The prime focus area is cardiology, where Implicity offers a solution to aggregate data from all cardiac electronic implantable devices across manufacturers. The platform also has features for optimizing patient scheduling and billing, alerting healthcare providers to clinical complications even before the patient becomes symptomatic, and easy integration with the patient's medical records. The raw structured data collected on the single platform also allow for designing clinical research programs to study the data further.
5. *Sutter Health* (Ekohealth 2021): Sutter Health has designed an efficient telehealth model to connect cardiologists to patients in rural counties in Northern California. Like Raid telehealth, Sutter is also an example of a clinic–clinic telemedicine solution where the patients are in the local clinic and the cardiologists are remote. The telehealth model utilizes a 3M Littmann core digital stethoscope (Ekohealth 2021), Eko platform, which has built-in AI algorithms to detect early signs of cardiopulmonary disease (Ekohealth 2021), Tryten telemedicine care (Tryten 2021), and the HIPAA-compliant Vidyo telehealth video conferencing service (Vidyo 2021).
6. *FibriCheck* (FibriCheck 2021): FibriCheck is a CE and FDA-cleared medical smartphone application to detect heart rhythm disorders. It uses the camera of a smartphone or the optical sensors of a smartwatch to detect heartbeats and derives a heart rhythm using a technique based on photoplethysmography. The data are sent to an online platform where predictive algorithms accurately diagnose the heart rhythm. FibriCheck has now devised a telehealth solution to support patients with AF during the COVID-19 pandemic.

The physician prescribes the app to the patients and requests them to measure the heart rhythm and rate for seven continuous days before a teleconsultation. The remote cardiologist then reviews the data to diagnose and recommend treatments.

4.4.3 Telepsychiatry/telemental health

There is a high prevalence of mental disorders in the world. However, specialized mental health services are highly deficient. To address this 'mental health gap', telepsychiatry has been instrumental in supporting psychiatric services across remote locations through the efficient use of information and communications technology (ICT). Since the 1960s there have been sporadic increases in technologies such as two-way television, telephones, etc, to provide emergency care for mental health patients (Wittson and Benschoter 1972, Dwyer 1973). The recent advancements in technologies have made mental health services more accessible, and experts can provide psychiatric consultation, assessment, and diagnosis to remote areas. Also, medication management and group psychotherapy could be facilitated through telepsychiatry. Telepsychiatry has also opened frontiers for storage and access to mental health data for research and educational purposes.

4.4.3.1 General framework

To deliver telepsychiatry services, providers and patients have used telephone calls, cell phone text messaging, emails, online chat forums, etc, for crisis management and consultation. Both synchronous and asynchronous delivery models have been used for diagnosis, symptom monitoring and management, psychotherapy, and providing psychiatric specialist support to primary care physicians in rural or underserved settings (Malhotra *et al* 2013). Video conferencing has been commonly used to deliver cognitive behavior therapy (Griffiths *et al* 2006).

Guided therapies facilitated by asynchronous telepsychiatry, text messaging, and smartphone apps are popular these days. They significantly reduce the need for human interaction, thus cutting costs, geospatial and time barriers. A recent study predicted the usefulness of automated therapy-based intelligent technologies and their capability to par with therapist-guided therapy for couple and family interventions (Doss *et al* 2017). Kiluk *et al* (2018) conducted a survey to compare the effectiveness of computerized cognitive behavioral therapy (CBT) administration versus clinician-delivered CBT for treating substance abuse disorders. The study concluded that patients using computerized CBT had significantly higher treatment retention rates, satisfaction, knowledge of CBT concepts, and reduced substance abuse frequency. Peer support-based apps such as PeerTECH have been developed to address the limited availability of psychiatrists. They have also proved to be feasible to be used by elderly patients (Fortuna *et al* 2018).

Usually, patients with mental health problems are reluctant to have human interaction and are more comfortable interacting with machines. In such cases, artificial intelligence could play a critical role in advancing mental health service delivery (Lucas *et al* 2014). Many apps use passive data to infer behavior from sensors and activities. Passive data is objective, quantifiable data received from sensors embedded in smartphones, smartwatches, and other mobile devices, and therefore does not require explicit user input. For example, in a study by Place *et al* (2017), parameters such as the total number of outgoing calls, location-based absolute distance traveled, voice modulations, speaking rate, and voice quality were captured. These parameters and clinical symptoms measured via interviews with a social worker were used to build AI-based predictive models to predict symptoms of depression and post traumatic stress disorder (PTSD).

Web-based CBT apps, although effective, are limited by poor adherence by users. Therefore, Fitzpatrick *et al* (2017) conducted a study to determine the effectiveness of a fully automated AI-based conversational agent called Woebot in providing a self-help program for college students with symptoms of anxiety and depression. Woebot used several machine learning and natural language processing techniques to converse with the user intelligently. The study demonstrated that the text-based conversational agent could provide an engaging alternative to conventional CBT. Artificial intelligence has empowered clinical decision making in recent times, making it possible to hold a conversation with a patient, undertake psychotherapy, and provide psychoeducation facilitated by various ICT tools (Luxton 2015).

4.4.3.2 Perceptions on telepsychiatry practice

Telepsychiatry is cost-effective and allows practitioners and patients to provide and receive care, irrespective of location and time. It also enables patients to receive specialized care that may otherwise not be available in their area. Telemental health solutions also reduce the number of hospital trips and allow for continuous monitoring. People who are not comfortable seeking in-person visits, such as those with anxiety or autism, could benefit significantly from telehealth solutions. It could decrease the stigma around mental healthcare. On the downside, providers may struggle to embrace the technology in their existing traditional approaches. Some mental health conditions do need human contact, and telehealth might not be suitable in those cases. The security and privacy of medical data should be carefully maintained using HIPAA-compliant platforms.

Telepsychiatry encompasses a variety of modes: messaging, video content, and mobile device data. There are various challenges involved with remote patient monitoring in telepsychiatry. First, only partial information can be perceived by the healthcare professionals (therapist, counselors, psychiatrist, etc) when using a telemedicine approach. So, in non-hospitalized patients, treatment has to be suggested based on the telehealth evaluation plus the recent appointment report. Second, there is a requirement for more immediate intervention if the symptoms change significantly in a short period, in particular in patients suffering from schizophrenia and depression. In these situations, a psychiatrist should perform continuous monitoring of the patient's symptoms to predict unprecedented emergency conditions. Technological advancements have enabled the effective collection and monitoring of health data even through smartphones. Since smartphones are available to most of the population, more mHealth intervention solutions and applications could be developed (Firth *et al* 2016).

The field of psychiatry is unique as human interaction is a critical element of its practice. As the evidence above demonstrates, these technologies are helpful to augment and extend and not replace psychiatric care. Engaging with and helping shape the course of these evolving fields of telepsychiatry will ensure that they continue to develop in a manner that benefits patients. Additionally, mental health professionals are uniquely positioned to ensure that these technologies respect the therapeutic relationship and remain rigorous in their scientific foundation.

4.4.3.3 Commercial applications

For therapists:

1. *Secure Telehealth* (Secure Telehealth 2021): Psychiatrists can use their computers and webcams to provide assessments and therapies through the Secure Telehealth app. There is also a provision to use two monitors. One shows high definition video, and the other shows the location on the medical records where the patient can type how they feel in real-time.
2. *SimplePractice* (SimplePractice 2021): This is another secure, HIPAA-compliant telehealth platform that integrates video calls. The platform also has options for scheduling, documentation, billing, and patient communication. It can help the psychiatrist engage with patients in real-time to share exercises, goal-tracking worksheets, etc.
3. *TheraPlatform* (TheraPlatform 2021): TheraPlatform is a HIPAA-compliant video conferencing and practice management software supporting teletherapy. There are options for group teletherapy for family, marriage, etc. The platform also has an integrated whiteboard that the providers can use for explaining concepts and allow the patients to perform therapeutic exercises such as painting, drawing, etc. There is also an option for the providers to observe the patient's experience of watching therapy-related videos. Their premier plan also offers interactive apps and games that could be useful for speech therapy, occupational therapy, etc.

For patients:

1. *Array Behavioral Care* (Array 2021): Array Behavioral Care is one of the USA's largest telepsychiatry service providers. It provides quality behavioral healthcare to both patients at home and clinicians at hospitals. Their AtHome service connects patients to licensed therapists and psychiatrists through video calls. For organizations, such as hospitals, clinics, primary care centers, correctional centers, employers, tribal communities, schools, etc, Array has options for the onsite care team to connect with Array's remote psychiatric professionals. They can evaluate patients via video calls and determine the next steps.
2. *Talkspace* (Talkspace 2021): Talkspace is a 24/7 text-based therapy service that provides options for individuals, couples, and teen therapies. Once the user takes a brief assessment, the app connects them to an appropriate therapist, who then takes the time to text message and provide therapy.

Talkspace also can allow users to get personalized treatment from an online psychiatrist and manage their prescriptions.

3. *Betterhelp* (Betterhelp 2021): Betterhelp is also a 24/7 text-based therapy service that allows the user to message their therapist anytime, anywhere without the need for scheduling. The users can also schedule a live session if needed.
4. *Online-Therapy* (Online-Therapy 2021): Online-Therapy is based on CBT, one of the most commonly used psychotherapeutic approaches for treating mental health problems. Once users sign up, they can access an online therapy program and have unlimited daily communications with a licensed therapist.
5. *Mental health apps*: Several smartphone-based apps have been developed to support users in handling mental health issues. These apps are reasonably priced, sometimes free, and offer a wealth of resources that are available 24/7. Many of these apps used AI to provide CBT and acceptance commitment therapy and address depression, eating disorders, anxiety, depression, and more (Gamble 2020).

4.4.4 Teledermatology

Healthcare delivery has been evolving since the advent of telemedicine. It can improve access to subspecialty expertise with minimum cost and improved quality of care. One such subspecialty is dermatology. Teledermatology is one of the most popular applications of telemedicine. In this field, skin images and associated clinical history are transferred to remote dermatologists through audio, visual, and data communication channels. Teledermatology facilitates the remote diagnosis of skin diseases by helping dermatologists interpret these skin images and clinical history. Lee *et al* (2018) provide a detailed overview of the recent studies that utilized teledermatology for diagnosing several types of skin disorders. For the last decade, the field of teledermatology has evolved by expanding access to remote places with lowered costs of care and improved healthcare monitoring systems.

4.4.4.1 General framework

For the framework, the asynchronous model, also called the 'store and forward' model, is the most commonly used one (Yim *et al* 2018). Digital skin images are captured from patients and sent to dermatologists for interpretation. This method is more cost-effective than the synchronous mode as there is no expensive video calling equipment needed for virtual patient-to-provider communication. In some situations, such as dermatopathology, real-time or interactive delivery models could use telerobotic microscopes to perform remote surgery.

Many organizations are now turning towards providing low-cost virtual care to patients rather than conventional face-to-face appointments to save time. Although the remote diagnoses are not par with the traditional in-person diagnoses, artificial intelligence tools and algorithms have begun to deliver promising results (Masood and Al-Jumaily 2013). Recently, studies have demonstrated that deep learning techniques could diagnose melanoma images with dermatologist-level performance (Esteva *et al* 2017, Winkler *et al* 2020). As AI in telemedicine is becoming an achievable reality, smartphone apps are also being developed as a means of teledermatology. Kochmann and Locatis (2016) reviewed all mobile teledermatology apps available at the Apple App Store. They observed that the apps varied in diagnostic scope, data collection methodologies, results (for the same patient condition), global coverage, and cost.

Teledermatology also includes sub-specialties such as teledermoscopy and teledermatopathology. In *teledermoscopy*, a microscope and digital camera are used to capture magnified digital images of skin lesions. These images are then transferred to a remote specialist for review and diagnosis (Börve *et al* 2013). Teledermoscopy is very beneficial in monitoring chronic skin conditions and optimizing treatments for the same. Since the technology has similar diagnostic accuracy in patients of all skin colors, it is non-discriminatory and globally deployable. The inclusion of both clinical and dermoscopic images has increased the diagnostic accuracy of teledermatology services (Uppal *et al* 2021). Image processing, machine learning, and deep learning techniques can also be used to classify skin lesions in dermoscopy images (Pathan *et al* 2018, Vasconcelos and Vasconcelos 2020).

In certain malignant and infectious skin conditions, clinical images alone are insufficient in accurately diagnosing skin lesions. Transfer of skin biopsy specimens could provide more information for an accurate diagnosis. For example, in the African Teledermatology Project (Tsang and Kovarik 2011), both clinical images and biopsy specimens were sent to the US to obtain a definitive diagnosis. In *teledermatopathology*, instead of sending physical skin biopsy samples, digitally scanned pathology slides of skin lesions are transferred to a remote pathologist who reviews the images on a computer screen to determine an assessment. With the advent of virtual slides, pathology slides could now be digitalized to higher resolution images for better remote analysis (Williams *et al* 2017). A study in Japan (Tsuchihashi 2011) indicated an improved adoption of virtual slides for educational purposes and consultation, diagnosis, and telepathology. Also, AI-based automated analysis of digitized pathology images could be adopted (Bera *et al* 2019). There are also options to use robotic telepathology (RT), where dermatologists remotely control the microscopes to adjust views and interpret the images (Fischer *et al* 2011, Thrall *et al* 2014). Another study proposed that smartphones could be attached to microscopes via adapters, and photomicrographs are captured using the smartphone's camera and sent via email to dermatologists (Lehman and Gibson 2013). Such studies open frontiers for smartphone devices to run AI computational algorithms on remote Cloud servers to support remote clinical decision support in teledermatology.

4.4.4.2 Perceptions of teledermatology practice

Approximately 23%–75% of the population, based on sociodemographic and disease-related factors, utilize the Internet for web-based symptom evaluation (Mueller *et al* 2017). Based on this response, there has been vast teledermatology practice in recent years (Yim *et al* 2018). Depending on location, specialist availability, and even insurance status, patients could access various screening, triaging, and diagnostic paths to maximize healthcare accessibility and affordability. AI-powered screening and diagnosis will likely help to expand the scope and variety of dermatologic care. AI-based deep learning methods could disrupt the field of teledermatology (Xiong *et al* 2019). However, further research is needed to determine how robust deep learning technologies will be, considering the inherent variability between patient images. Research in teledermatology should continue to refine deep learning methods to work with smartphone images and help physicians provide an effective remote diagnosis. Further, despite the rapid increase of teledermatology services, none of the services currently available offer reliable automated diagnostics. Another critical area of progress would be to develop safe, effective, and accessible automated predictive solutions that can provide valuable second opinions to patients and dermatologists.

Specific teledermatology challenges include operational inefficiency, lack of communication among providers and between patients and providers, and provider inexperience (Armstrong *et al* 2012). With technology advances, several companies offer provider-to-patient virtual consultation platforms and smartphone-based applications. Though more convenient and affordable to the patients, these modalities may need to incorporate proper privacy and security frameworks to ensure high-quality care delivery (Resneck *et al* 2016, Kochmann and Locatis 2016, Fogel and Sarin 2017). Over the past decade, there have been efforts to develop dermatological imaging standards, follow-up protocols, and national guidelines for teledermatology (Krupinski *et al* 2008, Eastman *et al* 2012). There have also been efforts to improve diagnostic technologies to address the above concerns.

4.4.4.3 Commercial applications

1. *SkinVision* (SkinVision 2021): SkinVision is a paid medical app that allows users to upload an image of a suspicious skin spot and uses a CE-approved, clinically validated algorithm to assess the image for the presence of malignancy. Based on the assessment, the app recommends continuous monitoring or scheduling an appointment with a dermatologist for further examination. The app has the potential to detect early signs of cancer if used regularly. Thissen *et al* (2017) evaluated the app for the risk assessment of skin lesions. The app uses fractal image analysis algorithms that were trained using 233 melanocytic and nonmelanocytic lesions. In 108 test images, the app demonstrated a sensitivity of 80% and a specificity of 78% in detecting pre-malignant lesions. In another study by Udrea *et al* (2020), an updated algorithm trained using 131,873 user-captured smartphone images, which were rated for risk by dermatologists, was validated. Validation was done on 285 validated malignant and 6000 validated benign cases. The algorithm presented a sensitivity of 95.1% and specificity of 78.3%.
2. *DermExpert* (VisualDX and DermExpert 2021): VisualDx is a diagnostic clinical decision support system that provides an exhaustive curated library of all types of medical images. It enables physicians to quickly search for diagnoses, drug reactions, and make a differential diagnosis. For

teledermatology, it has an AI-powered HIPAA-compliant solution called DermExpert that uses machine learning algorithms to analyze the skin image uploaded by the user.

3. *Melafind*: In 2011, the FDA-approved Melafind, a melanoma detection system that uses imaging processing to mark whether a suspicious pigmented skin lesion should be biopsied or not. In a recent observational study, 360 pigmented skin lesions were assessed by dermatologists using MelaFind and MelaFind demonstrated a high sensitivity (Monheit *et al* 2011, Fink *et al* 2017). Several other studies have also shown MelaFind's high sensitivity and how it could effectively reduce the number of unwanted biopsies (Shrivastava *et al* 2019). However, the product has been discontinued for sale since some of the software components were not included in the pre-market approval by the FDA.
4. *3Derm, now part of Digital Diagnostics* (3Derm 2021): 3Derm specializes in providing a teledermatology platform. It obtained two FDA Breakthrough Device designations for 3DermSpot, an AI-based algorithm to automatically detect several types of skin cancers.
5. *Other tools*: Advancements in artificial intelligence technologies with readily available web-based telehealth services have mitigated the problems related to automated diagnosis systems. Eventually, these advancements have given rise to the development of software-based dermatoscopic tools such as MoleMax (MoleMax 2021), FotoFinder (FotoFinder 2021), and FirstDerm (FirstDerm 2021). The advantages of these imaging modalities include reduced skin biopsy rates, lowered healthcare costs, and patient morbidity.

4.4.5 Teleophthalmology

Many people suffer from vision problems and there has been a surge in the requirement for ophthalmologists. Therefore, to address limited access to specialists and improve the ease and comfort in managing simple vision-related issues, teleophthalmology came into practice. In situations such as the COVID-19 pandemic, teleophthalmology has allowed ophthalmologists and patients to continue their care regimens (Kalavar *et al* 2020, Aguwa *et al* 2020). It has been a blessing for the early detection of vision-threatening eye diseases and providing access to quality vision care to patients in underprivileged and rural areas.

4.4.5.1 General framework

Screening and diagnosis can be performed in three ways: (i) using the apps installed in the patients' smartphones, (ii) sending trucks equipped with standard ophthalmological diagnostic equipment to the patients' localities, or (iii) using teleophthalmology platforms.

Ophthalmology is another medical field rich in medical images, and AI and image processing technologies are being harnessed for more accurate processing of these images (Li *et al* 2020). Recent years saw significant improvement in image recognition due to the advent of novel algorithms and techniques such as deep learning, and progress has been made in ophthalmic image segmentation, analysis, and clinical decision making. These AI-based algorithms can be embedded into teleophthalmology platforms and apps. Table 4.1 presents some developed AI algorithms that assist in diagnosing various visual disorders.

4.4.5.2 Perceptions of teleophthalmology practice

The COVID-19 pandemic initiated policy changes to allow for hospitals and healthcare providers to adopt telemedicine. However, in a study conducted at Johns Hopkins School of Medicine during the pandemic, it was observed that telemedicine adoption in the ophthalmology department was significantly lower than in other departments (Aguwa *et al* 2020). Female specialists and junior-level specialists were more likely to adopt telemedicine, and senior-level specialists, in particular cornea, glaucoma, and retinal specialists, were less likely to embrace telemedicine. Recommendations to encourage telemedicine adoption were as follows:

1. More technical training around teleophthalmology should be conducted to improve providers' confidence and interest.
2. Hybrid visits, a combination of at-home imaging using smartphone apps and in-clinic visits, could also encourage more telehealth service adoption. More AI-enabled clinical decision support algorithms could be embedded into home-based color fundus photography systems (Liu *et al* 2021).
3. Basic routine assessments and many examinations in sub-specialties such as oculoplastics, pediatric ophthalmology, and neuro-ophthalmology could be performed virtually. Therefore, these specialties could be more amenable to telehealth services (Saleem *et al* 2020).
4. Many ophthalmology patients are older, and simpler teleophthalmology platforms could improve the adoption rate among patients.

4.4.5.3 Commercial applications

Kortuem *et al* (2018) came up with a concept of virtual medical retina clinics in London. The virtual clinics were equipped to acquire color fundus photographs and OCT images. It was observed that the clinics could work as a first-line rapid-access option for low-risk patients, enabling proper management of increasing demand. Retina Labs (Retina Labs 2021) provides a cloud-based iVision teleophthalmology platform that uses advanced ocular imaging, clinical interpretation, and reporting tools for teleretinal screening of several types of visual disorders.

Recently, the FDA has been approving the application of AI for home-based diagnosis of diabetic retinopathy. Current research is more focused on integrating AI in point-of-care imaging devices and remote image-based diagnosis. Validated AI algorithms are augmenting virtual clinics to assist specialized ophthalmologists. The FDA-approved IDx-DR2.0 device is equipped with an AI-based automated DR grading algorithm (IDx-DR 2021). van der Heijden *et al* (2018) evaluated the device in a clinical setting and observed that it could be an efficient method to use primary

Table 4.1. Various ophthalmological diseases and AI-based decision support algorithms.

Ophthalmological disease	Examples of AI-based prediction/diagnostic algorithms
Glaucoma	• A deep learning algorithm to detect glaucomatous optic neuropathy based on color fundus photographs (Li *et al* 2018). • Machine learning models to classify glaucomatous eyes in optical coherence tomographic (OCT) scans (Kim *et al* 2017). • Deep learning models to classify glaucoma suspects in OCT scans (Muhammad *et al* 2017).
Cataracts	• A deep learning-based platform to manage congenital cataracts (Long *et al* 2017).
Ocular oncology	• Artificial neural networks to forecast the prognosis of metastatic choroidal melanoma (Kaiserman *et al* 2005). • A conditional hazard estimating neural network to predict the survival of patients with choroidal melanoma (Damato *et al* 2008).
Retinopathy	• Machine learning algorithms to automatically score avascular zones and segment vascular tufts (Mazzaferri *et al* 2018). • Deep neural networks to automate segmentation in oxygen induced retinopathy (OIR) images (Xiao *et al* 2017). • Machine learning-based image analysis system for grading plus disease in retinopathy of prematurity (Ataer-Cansizoglu *et al* 2015).
Age-related macular degeneration (AMD)	• A machine learning algorithm to grade AMD stages in OCT scans (Venhuizen *et al* 2017). • A deep learning algorithm to automatically grade AMD in color fundus images (Burlina *et al* 2017). • A predictive model to predict AMD progression using OCT scan images (Bogunovic *et al* 2017a). • A machine learning method to predict anti-vascular endothelial growth factor-based AMD treatment needs by analyzing OCT scans (Bogunovic *et al* 2017b).
Diabetic retinopathy (DR)	• A deep learning model to predict DR using color fundus images (Gargeya and Leng 2017). • A deep learning system to detect DR, glaucoma, and AMD in retinal images from a multiethnic cohort of patients with diabetes (Ting *et al* 2017). • A deep learning algorithm to grade DR in fundus photographs (Takahashi *et al* 2017). • A review of several deep learning algorithms in DR screening (Nielsen *et al* 2019).

care to help providers determine which patients should be referred to ophthalmologists. Hong *et al* (2019) observed promising results using a portable ophthalmic camera system powered by a mobile device for tele-opthalmic screening in low-resource areas. Kozak *et al* (2017) presented a model that implemented a retinal telephotocoagulation treatment for diabetic macular edema. In another study, Das *et al* (2019) observed that using the eyeSmart app and video calls proved beneficial in providing vision care to patients in rural areas.

4.4.6 Teleobstetrics

Telehealth is increasingly being used in women's healthcare, particularly in the following areas: virtual consultation, remote monitoring of ultrasound in maternal–fetal medicine, remote monitoring of conditions by endocrinologists, prenatal and postnatal monitoring, and fertility tracking (DeNicola *et al* 2020). Obstetric telemonitoring can improve gestational outcomes by providing local interventions to detect complications early on.

4.4.6.1 General framework

In teleobstetrics, communications can be both synchronous and asynchronous, just as in other applications of telemedicine. It can include virtual visits, remote patient monitoring, and mobile apps or text messaging. van den Heuvel *et al* (2018) reviewed publications on eHealth solutions for perinatal care in the areas of information delivery, lifestyle management, gestational diabetes, mental health, and telemedicine. In terms of telemonitoring, they summarized the development and evaluation of several remote monitoring systems for maternal care. These systems could track blood glucose levels, weight, blood pressure, fetal heart rate, and uterine contractions.

When integrated with AI, practitioners can use telemonitoring and obtain more reliable predictions, thus improving case management and finding an error-free diagnosis. Since AI can assist in the early detection of preterm labor and detect or differentiate various other pregnancy complications through the continuous monitoring of fetal heart rate and cardiotocography, infant morbidity and mortality can be reduced (Lee and Ahn 2020). Further, AI-based augmented reality has allowed surgeons to perform remote surgeries with more precision and has also acted as a reliable training platform for medical students. Iftikhar *et al* (2020) provide a detailed review on how AI can positively affect obstetrics and gynecology (OB/GYN). Telemonitoring integrated with AI can improve the prognosis, help patient management, reduce healthcare costs, minimize OB/GYN practitioners' workload, and increase diagnostic efficacy and accuracy. Researchers believe that AI technology can be used in smart home monitors to adequately monitor high-risk pregnancy patients in outpatient settings (Kazantsev *et al* 2012). These smart monitors combined with telemedicine can potentially help detect early signs of complications such as dangerous fetal heart rate readings. Since most families have access to the Internet and smartphones, outpatient monitoring can be performed easily.

4.4.6.2 Perceptions of teleobstetric practice

Studies have found that remote monitoring and smartphone apps helped decrease the number of unwanted outpatient visits to manage diabetes and hypertension in high-risk obstetric patients (DeNicola *et al* 2020). In general, studies have concluded that patient involvement and the availability of technologies for efficient home pregnancy care could lead to enhanced adoption of telehealth services in obstetrics (van den Heuvel *et al* 2018). As with any telehealth service, further improvements must be made to ensure patient privacy, security, and affordability. There is also a need to establish regulatory guidelines to provide a standard practice. Additional controlled clinical trials could help establish the reliability of remote monitoring in prenatal care (Lanssens *et al* 2017). Moreover, there is a need to develop and evaluate integrated obstetric telemonitoring systems that use various obstetric parameters for analysis (Alves *et al* 2020).

4.4.6.3 Commercial applications

1. *Babyscripts* (Babyscripts 2021): Babyscripts provides a virtual maternity care platform that allows physicians to deliver quality maternal care using a mobile app and a set of FDA-approved remote patient monitoring tools for blood pressure and weight (Mommy Kit). The app has a built-in machine learning-based trigger system that alerts the physicians to any elevated levels in the monitored parameters. The app also provides specific educational content and is facilitated by an intelligent chatbot regulated by natural language processing algorithms to promote effective care and take surveys on mental health and social risks.
2. *Rimidi* (Rimidi 2021): Rimidi's remote patient monitoring platform stores patient health data captured using connected devices such as blood pressure cuffs, weight scales, and blood glucose meters. It also collects and stores data from the patient's electronic health records to provide a comprehensive health outlook for continuous chronic disease management. Recently, Rimidi announced a partnership with Boston Medical Center to remotely monitor the blood pressure of women experiencing high-risk pregnancies (Rimidi and BMC 2021).
3. *Chiron* (Chiron 2021): Chiron allows OB/GYN practitioners to remotely manage medications for depression, anxiety, blood pressure, and birth control. It has a provision for remote video consultations to manage gestational diabetes and postpartum depression and discuss sudden hypertension and preeclampsia issues.
4. *Maven Clinic* (Maven 2021): Maven provides a virtual care model for women and families by providing a dedicated care advocate and options for on-demand appointments. The software also checks in with patients proactively to check for symptoms or send medication reminders.
5. *VisuWell* (VisuWell 2021): VisuWell provides a teleobstetric platform that can be integrated easily into the existing technology suite used by providers and hospitals. The platform has options for store and forward communications

and video sessions with specialists and is well suited to provide support to primary care physicians in rural areas.

4.4.7 Teleoncology

Teleoncology plays a pivotal role in providing affordable cancer care remotely to patients in underserved or rural areas. It is also beneficial to immunocompromised cancer patients during epidemics and pandemics (Shirke *et al* 2020).

4.4.7.1 General framework

Clinical oncology services can be delivered via audio and video conferencing technologies. The communication can happen synchronously or asynchronously or can even be hybrid combined with in-person care. In the event of the COVID-19 pandemic, the European Society for Medical Oncology (ESMO) recommended the use of phone and web-based toxicity evaluation, prescription dose management, and supportive care recommendation (Curigliano *et al* 2020). Telehealth services effectively provide remote chemotherapy supervision, symptom management, and palliative care to cancer patients. As discussed in a previous section, radiological services can be provided via teleradiology platforms. Innovations in technology have also enabled telepalliative care, which uses telehealth services such as phone and video conferencing, and remote monitoring to provide palliative and supportive care to patients with advanced cancer (Grewal *et al* 2020). Images, lab results, pathology reports, and even histology slides can be sent to specialists/pathologists to obtain a second opinion on the initial diagnosis using a telepathology consultation (He *et al* 2019). Wearable technologies and AI-enabled devices can be used for intermittent or continuous remote monitoring of signs and symptoms (Pennell *et al* 2017, Menta *et al* 2018).

4.4.7.2 Perceptions of teleoncology practice

Oncologists usually find the integration of teleoncology into their daily practices quite challenging and limited, as patients may also require a physical examination, biopsies, and surgeries. Although current technology can quickly enable remote patient interviews, consultations, imaging, and pathological examinations, the physical exam is still challenging. However, teleoncology services can still enhance existing oncology care (Hazin and Qaddoumi 2010). The shortage of oncologists, the aging population, and the need for specialty cancer care access in underserved communities and countries provide the rationale for developing, evaluating, implementing, and using teleoncology services. Further, an increasing distance of residence from specialty centers correlates with decreased utilization of high-quality care for cancers (Bristow *et al* 2014). In these cases, teleoncology apps and platforms could allow patients to adhere to their care protocol more effectively. Since cancer patients require care from a multidisciplinary team, teleoncology platforms that provide a bundle of services catering to all departments will be very beneficial.

4.4.7.3 Commercial applications

1. **Aurora** (Aurora 2021): Aurora provides an oncology-specific telemedicine platform that enables patients to consult physicians worldwide, access clinical trials and treatments available globally, and communicate easily with their oncologists.
2. **Biofourmis** (Biofourmis 2021): Biofourmis provides prescription software called Gaido to detect and monitor early signs of deterioration in cancer patients. Early detection helps prevent patients from developing more severe infections and reduces re-admissions.
3. **TempTraq** (Temptraq 2021): TempTraq is an FDA-cleared Class II medical device that detects, tracks, and sends body temperature data via Bluetooth to a TempTaq app installed on a smartphone (Apple or Android). TempTraq uses a soft, comfortable patch to measure temperature for up to 72 h. TempTraq Connect is a Google Healthcare Cloud Platform-supported HIPAA-compliant web service that allows the temperature information to be directly integrated with the hospital central monitoring systems. TempTraq has been observed to identify neutropenic fever much earlier in patients undergoing stem cell transplants or chemotherapy for leukemia (TempTraq and UH 2021).

4.4.8 Other applications of telemedicine

4.4.8.1 Telenephrology

Chronic kidney disease (CKD) is a global issue and is more prevalent due to the increasing incidence of diabetes and obesity. If left undetected or untreated, CKD could lead to the need for kidney replacement therapy, a costly procedure (Koraishy and Rohatgi 2020). A few barriers to delivering good renal care are as follows: (i) limited knowledge of CKD by primary care providers (PCP), (ii) limited availability of nephrologists, (iii) for-profit private health insurance, (iv) socioeconomic status and cultural or language barriers, and (v) lack of specialists closer to patients preventing earlier detection and diagnosis (Tan *et al* 2018).

In telenephrology, video conferencing is most commonly used to connect nephrologists and patients remotely. A technician on the patient's side could perform a standard examination under the remote nephrologist's guidance, who then recommends the steps based on the examination results. PCPs can also e-consult with nephrologists using the EHRs. Web-based applications and smartphone apps can also be used to effectively deliver educational content and allow patients to monitor vitals, and set up medication reminders and other health alerts. Alina Telehealth (Alina 2021), SOC Telemed (SOC 2021), and Sanderling (SRS 2021) are a few telenephrology service providers.

Reports from the Veterans Affairs in the US indicate that nephrology care delivery via telehealth services increases patient adherence to scheduled appointments (Rohatgi *et al* 2017). Video conferencing has also been commonly used in Australia to manage chronic kidney disease, dialysis, pediatric kidney care, and kidney transplantation care (Rohatgi *et al* 2017). Physicians must be trained to accurately identify those patients for whom telehealth technology could be beneficial (Crowley *et al* 2017).

4.4.8.2 Telerehabilitation

Telerehabilitation is the process of offering rehabilitation techniques to remote sites. It is used mainly to administer physiotherapy (Mani *et al* 2017), neurological telerehabilitation to monitor stroke patients (Jagos *et al* 2015), and cardiovascular telerehabilitation (Busch *et al* 2009). An example of cardiovascular rehabilitation is the SAPHIRE system (Busch *et al* 2009), which consists of an exercise bike equipped with a touch screen and wireless sensors to monitor ECG, blood pressure, and oxygen saturation during the use of the bike. Clinicians can connect remotely to the patient's bike computer to monitor and adapt the exercise programs to their needs. Telerehabilitation is a young field of telemedicine and further research is needed to establish long-term effectiveness and develop more reliable and effective telerehabilitation equipment (Peretti *et al* 2017). One example of a remote pain-management and online physical therapy platform is TheraNow (TheraNow 2021). NeuronUP (NeuronUP 2021) is an online platform that provides neuropsychological tools for neurorehabilitation and cognitive stimulation. The platform includes brain training exercises for Alzheimer's, Parkinson's, multiple sclerosis, etc.

4.4.8.3 Pediatric telehealth

Pediatric telehealth allows parents and pediatricians to work together to handle routine follow-ups for managing chronic conditions, mental and behavioral health and medication management, and sick visits for patients from newborns through to 21 years of age (Badawy and Radovic 2020). Grist *et al* (2019) conducted a systematic review and observed a medium effect size for Internet-based CBT for depression and anxiety in children and adolescents. Deacon and Edirippulige (2015) determined that self-management interventions for type 1 diabetes using mobile technologies would be more effective with physician support. In another study, it was found that remote monitoring and management of asthma through an electronic health intervention helped reduce the number of in-person visits (van den Wijngaart *et al* 2017).

Anytime Pediatrics (Anytime 2021) provides a pediatric telehealth platform that supports scheduling on-demand and virtual visits, automated smart routing to available pediatricians in the practice, billing, live chats, and the inclusion of educational resources. In addition to the usual pediatric conditions, Anytime also has features for virtual consultations for lactation, adoptions, and nutrition. Hazel is another telehealth provider, which focuses on providing telehealth services for K-12 schools (Hazel 2021). Chiron (Chiron 2021) also offers a platform for pediatric telehealth.

In pediatric behavioral health issues such as autism, attention disorders, and anxiety, early diagnosis and intervention during critical neurodevelopmental windows is key to improving outcomes for these children. Cognoa (Cognoa 2021) is developing clinically validated solutions for the early detection, diagnosis, and care of pediatric behavioral health issues. The company also recently received FDA

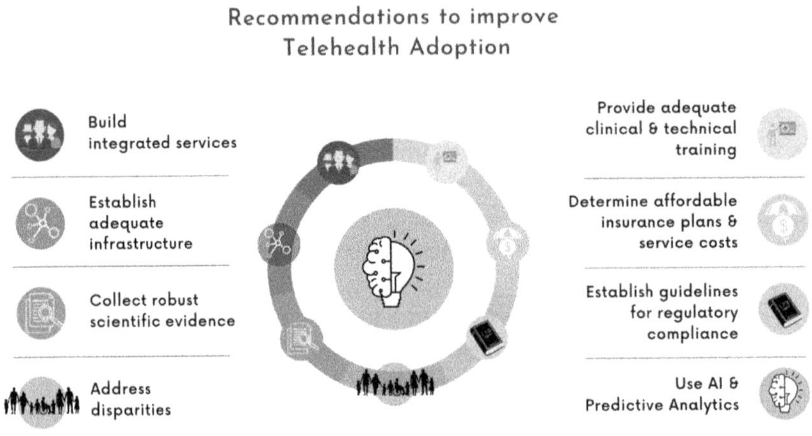

Recommendations to improve Telehealth Adoption

Build integrated services

Establish adequate infrastructure

Collect robust scientific evidence

Address disparities

Provide adequate clinical & technical training

Determine affordable insurance plans & service costs

Establish guidelines for regulatory compliance

Use AI & Predictive Analytics

Figure 4.2. Recommendations to improve telehealth adoption.

Breakthrough Device Designation for its digital diagnostic and therapeutic products for autism. Such digital solutions use parent-provided home-based video monitoring of kids and other parameters as inputs to be used in predictive algorithms to obtain a definitive diagnosis much earlier than when the condition is diagnosed in a clinical setting.

4.5 What is next?

This chapter provided an overview of the applications of telemedicine in different fields of medicine. Preliminary studies and architecture implementations show that a telemedicine approach is appropriate for the prevention, diagnosis, treatment, and short-term and long-term management of several health conditions. Specifically, telemedicine plays a critical role in providing quality and accessible healthcare to patients in low and middle-income countries and under-resourced regions in developed and developing nations (Bhaskar *et al* 2020a, 2020b). The COVID-19 pandemic has rapidly transformed healthcare systems worldwide and has led to significant variations in adaptation. For example, the US Government has temporarily waived several rules related to Health Insurance Portability and Accountability Act (HIPAA) regulations around telemedicine for audio and video communications (US HHS 2020). Additionally, insurers in the US have expanded their coverage and reimbursement of various home telehealth visits (CMS 2020).

The chapter provided a general idea of how telehealth services are enabled in radiology, cardiology, psychiatry, dermatology, ophthalmology, obstetrics, oncology, nephrology, rehabilitation, and pediatrics. Several commercial applications—platforms and applications—were also included. Each section also highlighted how telehealth was perceived by providers and patients and the areas that need further research. Below are a few of the general recommendations to improve telehealth adoption (figure 4.2):

1. *Integrated services*: A telehealth workflow process involves physicians and administrative support staff, nurses, social workers, case managers, patients,

and their families. Therefore, establishing evidence-based practices that integrate seamless communication between all involved parties is essential to address potential health risks. Moreover, the platforms and frameworks should also be effectively integrated with electronic medical records, electronic hospital records, smartphone apps, and remote monitoring devices. System interconnectivity between providers, pharmacies, laboratories, and hospital systems is also necessary to ensure appropriate care.

2. *Infrastructure*: Telehealth requires reliable broadband access, which is difficult for patients and clinics in rural areas. The widespread adoption of telehealth services should consider the availability of proper technological infrastructure, particularly in under-resourced settings.
3. *Address disparities*: Telehealth services are not always straightforward for patients with hearing, vision, and cognitive disabilities, nor are they comfortable for people belonging to ethnic minority groups due to possible language barriers. The services must incorporate solutions to address these disparities.
4. *Insurance and costs*: The risks and costs associated with the rapid adoption of new technologies should be adequately studied. Costs could also be affected by the type of disease and the digital infrastructure used for telehealth delivery. A standard reimbursement policy that is suitable for all stakeholders involved is necessary for sustainable adoption. America's Health Insurance Plans (AHIP), the national association of insurance providers in the US, has recently activated emergency plans in light of the COVID-19 pandemic. The plans detail how telemedicine services will be reimbursed by various insurance providers (AHIP 2021).
5. *Evidence*: Robust scientific analysis of existing research in telehealth should be conducted to inform providers and patients about the evidence to enable them to make informed decisions on harnessing telemedicine's benefits.
6. *Training*: Considerable training should be provided to both patients and providers to familiarize them with video teleconsultations and other supporting technologies. Physicians should also be provided with specialty-specific clinical, technical, and managerial training.
7. *Regulatory compliance*: There is an explosion in telemedicine-related legislation worldwide (Bhaskar *et al* 2020a). There is a need for standard legislation around cross-state licensing for physicians providing telehealth services across states, online prescribing, and informed consent from patients. Moreover, there should also be legal guidelines for telehealth platforms around patient data security and privacy (Hall and McGraw 2014). There should be a comprehensive and global telemedicine guideline for any country to adapt and customize based on nation-wise needs. This guideline should guide patient identification, data ownership, backup, data disposal, and cybersecurity laws (Intan Sabrina and Defi 2021).
8. *AI and predictive analytics*: AI-based algorithms for image processing, text mining, predictive modeling, wearable devices, and remote monitoring could play an essential role in providing telemedicine services. More research is

needed to determine robust algorithms that could be safely implemented into a telemedicine service. Bhaskar *et al* (2020c) proposed an AI-assisted telemedicine organizational framework to improve telehealth access across various medical specialties and global regions. The proposed AI system could automatically detect the best site of care for the patient's condition based on distance, availability of resources, clinicians' availability, and time constraints.

References

3Derm 2021 https://3derm.com/

ACC 2004 ACC/AHA guidelines for the management of patients with ST-elevation myocardial infarction *J. Am. Coll. Cardiol.* **44** 671–719

Ada 2021 https://ada.com/

Aguwa U T, Aguwa C J, Repka M, Srikumaran U, Woreta F, Singman E L, Jenkins S G and Srikumaran D 2020 Teleophthalmology in the era of COVID-19: characteristics of early adopters at a large academic institution *Telemed. J. E Health* **27** 739–46

AHIP 2021 https://ahip.org/health-insurance-providers-respond-to-coronavirus-covid-19/

Aldemir E, Gezer N S, Tohumoglu G, Barış M, Kavur A E, Dicle O and Selver M A 2020 Reversible 3D compression of segmented medical volumes: usability analysis for teleradiology and storage *Med. Phys.* **47** 1727–37

Alina 2021 https://alinatelehealth.com/telenephrology/

Alves D S, Times V C, da Silva É M A, Melo P S A and Novaes M A 2020 Advances in obstetric telemonitoring: a systematic review *Int. J. Med. Inform.* **134** 104004

Anytime 2021 https://anytimepediatrics.com/

Armstrong A W *et al* 2012 Teledermatology operational considerations, challenges, and benefits: the referring providers' perspective *Telemed. J. E Health* **18** 580–4

Array 2021 https://arraybc.com/organizations

Ataer-Cansizoglu E *et al* 2015 Computer-based image analysis for plus disease diagnosis in retinopathy of prematurity: performance of the 'i-ROP' system and image features associated with expert diagnosis *Transl. Vis. Sci. Technol.* **4** 5

Aurora 2021 https://aurorateleonco.com/en/about/

Babyscripts 2021 https://learn.babyscripts.com/home

Badawy S M and Radovic A 2020 Digital approaches to remote pediatric health care delivery during the COVID-19 pandemic: existing evidence and a call for further research *JMIR Pediatr. Parent.* **3** e20049

Bashshur R L, Howell J D, Krupinski E A, Harms K M, Bashshur N and Doarn C R 2016 The empirical foundations of telemedicine interventions in primary care *Telemed. J. E Health* **22** 342–75

Bera K, Schalper K A, Rimm D L, Velcheti V and Madabhushi A 2019 Artificial intelligence in digital pathology—new tools for diagnosis and precision oncology *Nat. Rev. Clin. Oncol.* **16** 703–15

Betterhelp 2021 https://betterhelp.com/

Bhaskar S *et al* 2020a Telemedicine across the globe-position paper from the COVID-19 pandemic health system resilience PROGRAM (REPROGRAM) int. consortium (Part 1) *Front. Public Health* **8** 556720

Bhaskar S *et al* 2020b Telemedicine as the new outpatient clinic gone digital: position paper from the pandemic health system REsilience PROGRAM (REPROGRAM) International Consortium (Part 2) *Front. Public Health* **8** 410

Bhaskar S *et al* 2020c Designing futuristic telemedicine using artificial intelligence and robotics in the COVID-19 era *Front. Public Health* **8** 556789

Biofourmis 2021 https://biofourmis.com/

Bogunovic H, Montuoro A, Baratsits M, Karantonis M G, Waldstein S M, Schlanitz F and Schmidt-Erfurth U 2017a Machine learning of the progression of intermediate age-related macular degeneration based on OCT imaging *Invest. Ophthalmol. Vis. Sci.* **58** BIO141–IO150

Bogunovic H, Waldstein S M, Schlegl T, Langs G, Sadeghipour A, Liu X, Gerendas B S, Osborne A and Schmidt-Erfurth U 2017b Prediction of anti-VEGF treatment requirements in neovascular AMD using a machine learning approach *Invest. Ophthalmol. Vis. Sci.* **58** 3240–8

Börve A, Terstappen K, Sandberg C and Paoli J 2013 Mobile teledermoscopy—there's an app for that! *Dermatol. Pract. Concept.* **3** 41–8

Bristow R E, Chang J, Ziogas A, Anton-Culver H and Vieira V M 2014 Spatial analysis of adherence to treatment guidelines for advanced-stage ovarian cancer and the impact of race and socioeconomic status *Gynecol. Oncol.* **134** 60–7

Burlina P M, Joshi N, Pekala M, Pacheco K D, Freund D E and Bressler N M 2017 Automated grading of age-related macular degeneration from color fundus images using deep convolutional neural networks *JAMA Ophthalmol.* **135** 1170–6

Busch C, Baumbach C, Willemsen D, Nee O, Gorath T, Hein A and Scheffold T 2009 Supervised training with wireless monitoring of ECG, blood pressure and oxygen-saturation in cardiac patients *J. Telemed. Telecare* **15** 112–4

Chiantera A *et al* 2005 Role of telecardiology in the assessment of angina in patients with recent acute coronary syndrome *J. Telemed. Telecare* **11** 93–4

Chiron 2021 https://chironhealth.com/telemedicine/providers/obstetrics-and-gynecology/

CMS 2020 Medicare Telemedicine Health Care Provider Fact Sheet (https://cms.gov/newsroom/fact-sheets/medicare-telemedicine-health-care-provider-fact-sheet)

Cognoa 2021 https://cognoa.com/

Crowley S T, Belcher J, Choudhury D, Griffin C, Pichler R, Robey B, Rohatgi R and Mielcarek B 2017 Targeting access to kidney care via telehealth: the VA experience *Adv. Chronic Kidney Dis.* **24** 22–30

Curigliano G *et al* 2020 Panel members. Managing cancer patients during the COVID-19 pandemic: an ESMO multidisciplinary expert consensus *Ann. Oncol.* **31** 1320–35

Damato B, Eleuteri A, Fisher A C, Coupland S E and Taktak A F 2008 Artificial neural networks estimating survival probability after treatment of choroidal melanoma *Ophthalmology.* **115** 1598–607

Das A V *et al* 2019 App-based tele ophthalmology: a novel method of rural eye care delivery connecting tertiary eye care center and vision centers in India *Int. J. Telemed. Appl.* **2019** 8107064

DeepTek 2021 https://deeptek.ai/teleradiology

Deacon A J and Edirippulige S 2015 Using mobile technology to motivate adolescents with type 1 diabetes mellitus: a systematic review of recent literature *J. Telemed. Telecare* **21** 431–8

De Bonis S, Salerno N, Bisignani A, Capristo A, Sosto G, Verta A, Borselli R, Capristo C and Bisignani G 2021 COVID-19 and STEMI: the role of telecardiology in the management of STEMI diagnosis during COVID 19 pandemic *Int. J. Cardiol. Heart Vasc.* **32** 100720

DeNicola N *et al* 2020 Telehealth interventions to improve obstetric and gynecologic health outcomes: a systematic review *Obstet. Gynecol.* **135** 371–82

De Ruvo E, Gargaro A, Sciarra L, De Luca L, Zuccaro L M, Stirpe F, Rebecchi M, Sette A, Lioy E and Calò L 2011 Early detection of adverse events with daily remote monitoring versus quarterly standard follow-up program in patients with CRT-D *Pacing Clin. Electrophysiol.* **34** 208–16

Doss B D, Feinberg L K, Rothman K, Roddy M K and Comer J S 2017 Using technology to enhance and expand interventions for couples and families: conceptual and methodological considerations *J. Fam. Psychol.* **31** 983–93

Dwyer T F 1973 Telepsychiatry: psychiatric consultation by interactive television *Am. J. Psych.* **130** 865–9

Eastman K L *et al* 2012 A teledermatology care management protocol for tracking completion of teledermatology recommendations *J. Telemed. Telecare* **18** 374–8

Ekohealth 2021 https://ekohealth.com/

Esteva A, Kuprel B, Novoa R A, Ko J, Swetter S M, Blau H M and Thrun S 2017 Dermatologist-level classification of skin cancer with deep neural networks *Nature* **542** 115–8

eVisit 2021 https://evisit.com/

FBI 2021 *Fortune Business Insights Report* (https://fortunebusinessinsights.com/industry-reports/telehealth-market-101065)

FibriCheck 2021 https://fibricheck.com/telecheck-af

Fink C, Jaeger C, Jaeger K and Haenssle H A 2017 Diagnostic performance of the MelaFind device in a real-life clinical setting *J. Dtsch Dermatol. Ges.* **15** 414–9

FirstDerm 2021 https://firstderm.com/

Firth J, Cotter J, Torous J, Bucci S, Firth J A and Yung A R 2016 Mobile phone ownership and endorsement of 'mHealth' among people with psychosis: a meta-analysis of cross-sectional studies *Schizophr. Bull.* **42** 448–55

Fischer M K, Kayembe M K, Scheer A J, Introcaso C E, Binder S W and Kovarik C L 2011 Establishing telepathology in Africa: lessons from Botswana *J. Am. Acad. Dermatol.* **64** 986–7

Fitzpatrick K K, Darcy A and Vierhile M 2017 Delivering cognitive behavior therapy to young adults with symptoms of depression and anxiety using a fully automated conversational agent (Woebot): a randomized controlled trial *JMIR Ment. Health.* **4** e19

Fogel A L and Sarin K Y 2017 A survey of direct-to-consumer teledermatology services available to US patients: explosive growth, opportunities and controversy *J. Telemed. Telecare* **23** 19–25

Fortuna K L, DiMilia P R, Lohman M C, Bruce M L, Zubritsky C D, Halaby M R, Walker R M, Brooks J M and Bartels S J 2018 Feasibility, acceptability, and preliminary effectiveness of a peer-delivered and technology supported self-management intervention for older adults with serious mental illness *Psychiatr. Q.* **89** 293–305

FotoFinder 2021 https://fotofinder-systems.com/technology/skin-cancer-screening/

Gamble A 2020 Artificial intelligence and mobile apps for mental healthcare: a social informatics perspective *Aslib J. Inf. Manag.* **72** 509–23

Gargeya R and Leng T 2017 Automated identification of diabetic retinopathy using deep learning *Ophthalmology.* **124** 962–9

Georgiadis P, Cavouras D, Daskalakis A, Sifaki K, Malamas M, Nikiforidis G and Solomou E 2007 PDA-based system with teleradiology and image analysis capabilities *Annu. Int. Conf. IEEE Eng. Med. Biol. Soc.* **2007** 3090–3

Grewal U S, Terauchi S and Beg M S 2020 Telehealth and palliative care for patients with cancer: implications of the COVID-19 pandemic *JMIR Cancer.* **6** e20288

Griffiths L, Blignault I and Yellowlees P 2006 Telemedicine as a means of delivering cognitive-behavioural therapy to rural and remote mental health clients *J. Telemed. Telecare* **12** 136–40

Grist R, Croker A, Denne M and Stallard P 2019 Technology delivered interventions for depression and anxiety in children and adolescents: a systematic review and meta-analysis *Clin. Child Fam. Psychol. Rev.* **22** 147–71

Hall J L and McGraw D 2014 For telehealth to succeed, privacy and security risks must be identified and addressed *Health Aff. (Millwood)* **33** 216–21

Hazel 2021 https://hazel.co/

Hazin R and Qaddoumi I 2010 Teleoncology: current and future applications for improving cancer care globally *Lancet Oncol.* **11** 204–10

He G *et al* 2019 Experience and impacts of remote telepathology consultation in improving cancer patients' diagnosis and management in China *J. Clin. Oncol.* **37** e18337

Healthtap 2021 https://healthtap.com/

Hjorth-Hansen A K, Andersen G N, Graven T, Gundersen G H, Kleinau J O, Mjølstad O C, Skjetne K, Stølen S, Torp H and Dalen H 2020 Feasibility and accuracy of tele-echocardiography, with examinations by nurses and interpretation by an expert via tele-medicine, in an outpatient heart failure clinic *J. Ultrasound Med.* **39** 2313–23

Hong K *et al* 2019 Teleophthalmology through handheld mobile devices: a pilot study in rural Nepal *J. Mob. Technol. Med.* **8**

Hsieh J C and Lo H C 2010 The clinical application of a PACS-dependent 12-lead ECG and image information system in e-medicine and telemedicine *J. Digit. Imaging.* **23** 501–13

IDx-DR 2021 https://digitaldiagnostics.com/products/eye-disease/idx-dr/

Iftikhar P, Kuijpers M V, Khayyat A, Iftikhar A and DeGouvia De Sa M 2020 Artificial intelligence: a new paradigm in obstetrics and gynecology research and clinical practice *Cureus* **12** e7124

Implicity 2021 https://implicity.com/our-solutions/cardiac-remote-monitoring-platform/

Intan Sabrina M and Defi I R 2021 Telemedicine guidelines in South East Asia—a scoping review *Front. Neurol.* **11** 581649

InTouch 2021 https://intouchhealth.com/virtual-care-platform/solo/

Insulander P, Carnlöf C, Schenck-Gustafsson K and Jensen-Urstad M 2020 Device profile of the Coala Heart Monitor for remote monitoring of the heart rhythm: overview of its efficacy *Expert Rev. Med. Devices.* **17** 159–65

Jackson S L, Tong X, King R J, Loustalot F, Hong Y and Ritchey M D 2018 National burden of heart failure events in the United States, 2006 to 2014 *Circ. Heart. Fail.* **11** e004873

Jagos H *et al* 2015 A framework for (tele-) monitoring of the rehabilitation progress in stroke patients *Appl. Clin. Inform.* **6** 757–68

Kaiserman I, Rosner M and Pe'er J 2005 Forecasting the prognosis of choroidal melanoma with an artificial neural network *Ophthalmology* **112** 1608

Kalavar M, Hua H U and Sridhar J 2020 Teleophthalmology: an essential tool in the era of the novel coronavirus 2019 *Curr. Opin. Ophthalmol.* **31** 366–73

Kazantsev A, Ponomareva J, Kazantsev P, Digilov R and Huang P 2012 Development of e-health network for in-home pregnancy surveillance based on artificial intelligence *Proc. IEEE-EMBS Int. Conf. on Biomedical and Health Informatics: Global Grand Challenge of Health Informatics* pp 82–4

Kiluk B D, Nich C, Buck M B, Devore K A, Frankforter T L, LaPaglia D M, Muvvala S B and Carroll K M 2018 Randomized clinical trial of computerized and clinician-delivered CBT in comparison with standard outpatient treatment for substance use disorders: primary within-treatment and follow-up outcomes *Am. J. Psych.* **175** 853–63

Kim S J, Cho K J and Oh S 2017 Development of machine learning models for diagnosis of glaucoma *PLoS One* **12** e0177726

Kochmann M and Locatis C 2016 Direct to consumer mobile teledermatology apps: an exploratory study *Telemed. J. E-Health* **22** 689–93

Koraishy F M and Rohatgi R 2020 Telenephrology: an emerging platform for delivering renal health care *Am. J. Kidney Dis.* **76** 417–26

Kortuem K *et al* 2018 Implementation of medical retina virtual clinics in a tertiary eye care referral centre *Br. J. Ophthalmol.* **102** 1391–5

Kozak I, Payne J F, Schatz P, Al-Kahtani E and Winkler M 2017 Teleophthalmology image-based navigated retinal laser therapy for diabetic macular edema: a concept of retinal telephotocoagulation *Graefes Arch. Clin. Exp. Ophthalmol.* **255** 1509–13

Krishnan A, Fuska M, Dixon R and Sable C A 2014 The evolution of pediatric tele-echocardiography: 15-year experience of over 10,000 transmissions *Telemed. J. E Health.* **20** 681–6

Krupinski E *et al* 2008 American Telemedicine Association's practice guidelines for teledermatology *Telemed. J. E Health* **14** 289–302

Kuziemsky C *et al* 2019 Role of artificial intelligence within the telehealth domain *Yearb. Med. Inform.* **28** 35–40

Lanssens D, Vandenberk T, Thijs I M, Grieten L and Gyselaers W 2017 Effectiveness of telemonitoring in obstetrics: scoping review *J. Med. Internet Res.* **19** e327

Lee K J, Finnane A and Soyer H P 2018 Recent trends in teledermatology and teledermoscopy *Dermatol. Pract. Concept.* **8** 214–23

Lee K S and Ahn K H 2020 Application of artificial intelligence in early diagnosis of spontaneous preterm labor and birth *Diagnostics* **10** 733

Lehman J S and Gibson L E 2013 Smart teledermatopathology: a feasibility study of novel, high-value, portable, widely accessible and intuitive telepathology methods using handheld electronic devices *J. Cutan. Pathol.* **40** 513–8

Lemonaid Health 2021 https://lemonaidhealth.com/

Li Z, He Y, Keel S, Meng W, Chang R T and He M 2018 Efficacy of a deep learning system for detecting glaucomatous optic neuropathy based on color fundus photographs *Ophthalmology* **125** 1199–206

Li J O *et al* 2020 Digital technology, tele-medicine and artificial intelligence in ophthalmology: a global perspective *Prog. Retin. Eye Res.* **82** 100900

Liu T Y A, Hui F K and Phan P H 2021 Clearing hurdles to large scale telemedicine adoption *Ophthalmology Times* (www.ophthalmologytimes.com/view/clearing-hurdles-to-large-scale-telemedicine-adoption?eKey=ZHNyaWt1bWFyYW5AZ21haWwuY29t)

Long E *et al* 2017 An artificial intelligence platform for the multihospital collaborative management of congenital cataracts *Nat. Biomed. Eng.* **1** 24

Lucas G M, Gratch J, King A and Morency L P 2014 It's only a computer: virtual humans increase willingness to disclose *Comput. Human Behav.* **37** 94–100

Luxton D D 2015 *Artificial Intelligence in Behavioral and Mental Health Care* ed D David (New York: Academic)

Mabo P, Victor F, Bazin P, Ahres S, Babuty D, Da Costa A, Binet D and Daubert J C 2012 A randomized trial of long-term remote monitoring of pacemaker recipients (the COMPAS trial) *Eur. Heart J.* **33** 1105–11

Magnusson P, Lyren A and Mattsson G 2020 Diagnostic yield of chest and thumb ECG after cryptogenic stroke, Transient ECG Assessment in Stroke Evaluation (TEASE): an observational trial *BMJ Open.* **10** e037573

Malhotra S, Chakrabarti S and Shah R 2013 Telepsychiatry: promise, potential, and challenges *Indian J. Psych.* **55** 3–11

Mani S, Sharma S, Omar B, Paungmali A and Joseph L 2017 Validity and reliability of Internet-based physiotherapy assessment for musculoskeletal disorders: a systematic review *J. Telemed. Telecare* **23** 379–91

Masood A and Al-Jumaily A A 2013 Computer-aided diagnostic support system for skin cancer: a review of techniques and algorithms *Int. J. Biomed. Imaging* **2013** 323268

Maven 2021 https://mavenclinic.com/

Mazzaferri J, Larrivée B, Cakir B, Sapieha P and Costantino S 2018 A machine learning approach for automated assessment of retinal vasculature in the oxygen induced retinopathy model *Sci. Rep.* **2018** 3916

Medweb 2021 https://medweb.com/

Menta A K, Subbiah I M and Subbiah V 2018 Bringing wearable devices into oncology practice: fitting smart technology in the clinic *Discov. Med.* **26** 261–70

MILLENSYS 2021 https://millensys.com/products/telerad/index.html

MoleMax 2021 https://dermamedicalsystems.us/molemax-systems

Molinari G, Valbusa A, Terrizzano M, Bazzano M, Torelli L, Girardi N and Barsotti A 2004 Nine years' experience of telecardiology in primary care *J. Telemed. Telecare* **10** 249–53

Molinari G, Molinari M, Di Biase M and Brunetti N D 2018 Telecardiology and its settings of application: an update *J. Telemed. Telecare* **24** 373–81

Monheit G *et al* 2011 The performance of MelaFind: a prospective multicenter study *Arch. Dermatol.* **147** 188–94

Mueller J, Jay C, Harper S, Davies A, Vega J and Todd C 2017 Web use for symptom appraisal of physical health conditions: a systematic review *J. Med. Internet Res.* **19** e202

Muhammad H, Fuchs T J, De Cuir N, De Moraes C G, Blumberg D M, Liebmann J M, Ritch R and Hood D C 2017 Hybrid deep learning on single wide-field optical coherence tomography scans accurately classifies glaucoma suspects *J. Glaucoma* **26** 1086–94

NeuronUP 2021 https://neuronup.com/

Nielsen K B, Lautrup M L, Andersen J K H, Savarimuthu T R and Grauslund J 2019 Deep learning-based algorithms in screening of diabetic retinopathy: a systematic review of diagnostic performance *Ophthalmol. Retina.* **3** 294–304

Nines 2021 https://nines.com

Ohmni 2021 https://ohmnilabs.com/products/ohmnirobot/

Online-Therapy 2021 https://online-therapy.com/how_it_works.php

Orbis 2021 https://orbis.org/en

Pathan S, Prabhu K G and Siddalingaswamy P C 2018 Techniques and algorithms for computer-aided diagnosis of pigmented skin lesions—a review *Biomed. Signal Process. Control* **39** 237–62

Paxera Health 2021 https://paxerahealth.com/products/pacs-system/teleradiology/

Pennell N A, Dicker A P, Tran C, Jim H S L, Schwartz D L and Stepanski E J 2017 mHealth: mobile technologies to virtually bring the patient into an oncology practice *Am. Soc. Clin. Oncol. Educ.* B **37** 144–54

Peretti A, Amenta F, Tayebati S K, Nittari G and Mahdi S S 2017 Telerehabilitation: review of the state-of-the-art and areas of application *JMIR Rehabil. Assist. Technol.* **4** e7

Place S *et al* 2017 Behavioral indicators on a mobile sensing platform predict clinically validated psychiatric symptoms of mood and anxiety disorders *J. Med. Internet Res.* **19** e75

Reid 2021 https://reidhealth.org/services/telecardiology

Resneck J S Jr *et al* 2016 Choice, transparency, coordination, and quality among direct-to-consumer telemedicine websites and apps treating skin disease *JAMA Dermatol.* **152** 768–75

Retina Labs 2021 https://retina-labs.com/company/

Rimidi 2021 https://rimidi.com/solutions

Rimidi and BMC 2021 https://healthitoutcomes.com/doc/boston-medical-center-partners-rimidi-partum-remote-monitoring-pregnancies-0001

Rohatgi R, Ross M J and Majoni S W 2017 Telenephrology: current perspectives and future directions *Kidney Int.* **92** 1328–33

Rosenkrantz A B, Hanna T N, Steenburg S D, Tarrant M J, Pyatt R S and Friedberg E B 2019 The current state of teleradiology across the United States: a national survey of radiologists' habits, attitudes, and perceptions on teleradiology practice *J. Am. Coll. Radiol.* **16** 1677–87

Roth A, Malov N, Steinberg D M, Yanay Y, Elizur M, Tamari M and Golovner M 2009 Telemedicine for post-myocardial infarction patients: an observational study *Telemed. J. E Health* **15** 24–30

Saberian P, Tavakoli N, Ramim T, Hasani-Sharamin P, Shams E and Baratloo A 2019 The role of pre-hospital telecardiology in reducing the coronary reperfusion time; a brief report *Arch. Acad. Emerg. Med.* **7** e15

Saleem S M, Pasquale L R, Sidoti P A and Tsai J C 2020 Virtual ophthalmology: telemedicine in a COVID-19 era *Am. J. Ophthalmol.* **216** 237–42

Scalvini S, Zanelli E, Comini L, Tomba M D, Troise G and Giordano A 2009 Home-based exercise rehabilitation with telemedicine following cardiac surgery *J. Telemed. Telecare* **15** 297–301

Schwaab B *et al* 2005 Pre-hospital diagnosis of myocardial ischaemia by telecardiology: safety and efficacy of a 12-lead electrocardiogram, recorded and transmitted by the patient *J. Telemed. Telecare* **11** 41–4

Secure Telehealth 2021 https://securetelehealth.com/

SkinVision 2021 https://skinvision.com/

Shen Y *et al* 2021 Robots under COVID-19 pandemic: a comprehensive survey *IEEE Access.* **9** 1590–615

Shirke M M, Shaikh S A and Harky A 2020 Tele-oncology in the COVID-19 era: the way forward? *Trends Cancer* **6** 547–49

Shrivastava V *et al* 2019 Histopathologic correlation of high-risk MelaFind^{TM} lesions: a 3-year experience from a high-risk pigmented lesion clinic *Int. J. Dermatol.* **58** 569–76

Simple Practice 2021 https://simplepractice.com/telehealth/

SOC 2021 https://soctelemed.com/telenephrology/

SRS 2021 https://srs-usa.com/telenephrologists/

Steventon A *et al* 2012 Effect of telehealth on use of secondary care and mortality: findings from the whole system demonstrator cluster randomised trial *Brit. Med. J.* **344** e3874

Takahashi H, Tampo H, Arai Y, Inoue Y and Kawashima H 2017 Applying artificial intelligence to disease staging: deep learning for improved staging of diabetic retinopathy *PLoS One.* **12** e0179790

Talkspace 2021 https://talkspace.com/

Tan J, Mehrotra A, Nadkarni G N, He J C, Langhoff E, Post J, Galvao-Sobrinho C, Thode H C Jr and Rohatgi R 2018 Telenephrology: providing healthcare to remotely located patients with chronic kidney disease *Am. J. Nephrol.* **47** 200–7

TempTraq 2021 https://bluesparktechnologies.com/

TempTraq and UH 2021 https://uhhospitals.org/for-clinicians/articles-and-news/articles/2017/06/clinical-study-shows-temptraq-wearable-bluetooth-continuous-temperature-monitor

TheraNow 2021 https://theranow.com/

TheraPlatform 2021 https://theraplatform.com/features/hipaa-compliant-video-conferencing

Thissen M, Udrea A, Hacking M, von Braunmuehl T and Ruzicka T 2017 mHealth app for risk assessment of pigmented and nonpigmented skin lesions—a study on sensitivity and specificity in detecting malignancy *Telemed. J. E Health* **23** 948–54

Thrall M J, Rivera A L, Takei H and Powell S Z 2014 Validation of a novel robotic telepathology platform for neuropathology intraoperative touch preparations *J. Pathol. Inform.* **5** 21

Ting D S W *et al* 2017 Development and validation of a deep learning system for diabetic retinopathy and related eye diseases using retinal images from multiethnic populations with diabetes *JAMA* **318** 2211–23

Tryten 2021 https://tryten.com/

Tsang M W and Kovarik C L 2011 The role of dermatopathology in conjunction with teledermatology in resource-limited settings: lessons from the African Teledermatology Project *Int. J. Dermatol.* **50** 150–6

Tsuchihashi Y 2011 Expanding application of digital pathology in Japan—from education, telepathology to autodiagnosis *Diagn. Pathol.* **6** S19

Udrea A, Mitra G D, Costea D, Noels E C, Wakkee M, Siegel D M, de Carvalho T M and Nijsten T E C 2020 Accuracy of a smartphone application for triage of skin lesions based on machine learning algorithms *J. Eur. Acad. Dermatol. Venereol.* **34** 648–55

Uppal S K, Beer J, Hadeler E, Gitlow H and Nouri K 2021 The clinical utility of teledermoscopy in the era of telemedicine *Dermatol. Ther.* **34** e14766

US HHS 2020 Notification of enforcement discretion for telehealth remote communications during the COVID-19 nationwide public health emergency (https://hhs.gov/hipaa/for-professionals/special-topics/emergency-preparedness/notification-enforcement-discretion-telehealth/index.html)

van der Heijden A A, Abramoff M D, Verbraak F, van Hecke M V, Liem A and Nijpels G 2018 Validation of automated screening for referable diabetic retinopathy with the IDx-DR device in the Hoorn Diabetes Care System *Acta Ophthalmol.* **96** 63–8

van den Heuvel J F *et al* 2018 eHealth as the next-generation perinatal care: an overview of the literature *J. Med. Internet Res.* **20** e202 van den Wijngaart L S *et al* 2017 A virtual asthma clinic for children: fewer routine outpatient visits, same asthma control *Eur. Respir. J.* **50** 1700471

Vasconcelos C N and Vasconcelos B N 2020 Experiments using deep learning for dermoscopy image analysis *Pattern Recognit. Lett.* **139** 95–103

Venhuizen F G, van Ginneken B, van Asten F, van Grinsven M J J P, Fauser S, Hoyng C B, Theelen T and Sánchez C I 2017 Automated staging of age-related macular degeneration using optical coherence tomography *Invest. Ophthalmol. Vis. Sci.* **58** 2318–28

Vidyo 2021 https://vidyo.com/video-conferencing-solutions/industry/telehealth

VisualDX and DermaExpert 2021 https://visualdx.com/

VisuWell 2021 https://visuwell.io/telemedicine/obgyn-teleobstetrics/

WHO 2012 https://paho.org/hq/index.php?option=com_content&view=article&id=7410:2012-dia-radiografia-dos-tercios-poblacion-mundial-no-tiene-acceso-diagnostico-imagen&Itemid=1926&lang=en

Williams B J, Bottoms D and Treanor D 2017 Future-proofing pathology: the case for clinical adoption of digital pathology *J. Clin. Pathol.* **70** 1010–8

Winkler J K *et al* 2020 Melanoma recognition by a deep learning convolutional neural network-performance in different melanoma subtypes and localisations *Eur. J. Cancer.* **127** 21–9

Wittson C L and Benschoter R 1972 Two-way television: helping the Medical Center reach out *Am. J. Psych.* **129** 624–7

Xiao S *et al* 2017 Fully automated, deep learning segmentation of oxygen-induced retinopathy images *JCI Insight.* **2** e97585

Xiong M, Pfau J, Young A T and Wei M L 2019 Artificial intelligence in teledermatology *Curr. Dermatol. Rep.* **8** 85–90

Yim K M, Florek A G, Oh D H, McKoy K and Armstrong A W 2018 Teledermatology in the United States: an update in a dynamic era *Telemed. J. E Health* **24** 691–7

IOP Publishing

Predictive Analytics in Healthcare, Volume 1
Transforming the future of medicine
Vinithasree Subbhuraam

Chapter 5

Pervasive healthcare applications in neurology

Vinithasree Subbhuraam and Dyuti Kumar

Pervasive healthcare is the concept of providing healthcare to anyone, anywhere, and anytime. The existence and use of good quality wireless and mobile network infrastructure have supported the growth of pervasive healthcare applications. These applications primarily include solutions for symptom management, incidence detection, and emergency intervention via short-term monitoring (home healthcare) and long-term monitoring (nursing home care). This chapter provides an overview of some of the pervasive solutions to manage the two most common neurocognitive disorders, namely Alzheimer's disease (AD) and Parkinson's disease (PD), and the most common neurological disorder, epilepsy. Successful management of these disorders is possible through continuous monitoring of symptoms and timely interventions, and wearables and remote monitoring devices are apt for this purpose. The chapter also highlights how predictive analytic algorithms play a critical role in the use of pervasive devices. Finally, the challenges and advancements that are necessary for the technology and application areas are discussed.

5.1 Introduction

Neurological disorders are conditions related to abnormal functioning of the brain, spine, and the nerves that connect them (the central and peripheral nervous systems). Some of the most common neurological disorders are epilepsy, neurocognitive disorders such as Alzheimer's disease (AD) and Parkinson's disease (PD), multiple sclerosis, brain tumors, strokes, migraines, etc.

Pervasive healthcare is a term coined to describe quality healthcare available to anyone worldwide irrespective of location and time zone. The development and advancements in technologies such as wearable sensors, remote monitoring devices, secure data transmission protocols, high-speed Internet connections, the Cloud, and practical and safe frameworks integrating all data collection components, transmission, storage, and retrieval and analysis have enabled pervasive healthcare. Furthermore, the advancements in data analysis methodologies such as machine

doi:10.1088/978-0-7503-2312-3ch5

learning, deep learning, predictive analytics, computer vision, image and text processing, etc, have also been beneficial to the evolution of pervasive devices. Pervasive healthcare applications include early detection and symptom management by remote health monitoring at home and long-term centers such as nursing homes, emergency incidence detection, alerting, and treatment monitoring.

In managing most chronic diseases, continuous monitoring of symptoms is generally more effective than occasional episodic reviews at clinics. This chapter provides an overview of some of the pervasive devices developed to manage the two most common neurocognitive disorders, namely AD and PD, and the most common neurological disorder, epilepsy.

5.2 Alzheimer's disease

AD is a progressive and irreversible neurological disorder that affects memory, thinking ability, and behavior. AD is the sixth leading cause of morbidity in the United States, with 122 019 deaths reported officially in 2018 (Alzheimer's Association 2020). It is the most common form of dementia observed in older individuals, with an estimated 5.8 million of the American population being affected (Alzheimer's Association 2020). The pre-clinical stage, which can last for several years, is generally asymptomatic but involves the beginning of deposition of the proteins, beta-amyloid plaques, and neurofibrillary tangles of hyperphosphorylated tau across the brain. The functioning neurons are eventually unable to function, thereby leading to apoptosis of once-healthy neurons. Initially, the hippocampus and medial temporal lobe, both involved in forming memories, are damaged, resulting in impaired memory forming. The brain gradually shrinks as the damage spreads until the brain tissue shrinks significantly, indicating the final stage of AD. AD can be classified into three main stages (Maresova *et al* 2018), as shown in table 5.1.

Current treatments for AD can neither reverse the condition nor slow down its progression. However, they can decelerate dementia symptoms for a short time (Cova *et al* 2017). Additionally, studies indicate that the early identification of individuals at risk and, consequently, treatment in the pre-clinical stages could allow for better management of the disease (Doraiswamy *et al* 2018). Thus, early detection could be beneficial for the person with AD and the caregiver. In 2018, family members and unpaid help totaled 16 million, with approximately 18.5 billion hours spent caring for AD patients (Alzheimer's Association 2020). Caregivers can also become emotionally, mentally, and physically distressed and exhausted. The inclusion of pervasive devices can promote the patient's autonomy in their place of care and can prove to be a stimulus in fostering the social integration of individuals with AD. Such devices can also reduce the load on caretakers while also being more economical for AD.

5.2.1 Pervasive devices for AD

There is a vast development of pervasive technologies for AD management in the areas of cognitive screening (through mobile EEG, speech recognition, and tablet-based tests), motor function monitoring (such as gait assessment), and behavior

Table 5.1. Various stages of AD.

Stage	Autonomy	Apparentness of symptoms	Symptoms
Mild (early)	May function independently	Not very	Difficulty in remembering new names or correct words/names, misplacing valuables, trouble organizing, and trouble remembering what was just read.
Moderate (middle)	May require assistance	Somewhat	Forgetfulness, moodiness, inability to recall personal information (address, phone number), confusion about orientation, bladder and bowel control issue, change in sleep pattern marked with restlessness at night.
Severe (late)	Complete dependence on a caretaker	Very	Loss of awareness, communication problems, vulnerability to infections, change in physical abilities (walking, sitting, and eventually swallowing).

monitoring (using cameras to assess mood, track movements, and facial emotions, Global Positioning System (GPS) trackers to monitor wandering behaviors). Doraiswamy *et al* (2018) provide a detailed overview of these technologies and applications. Kourtis *et al* (2019) also review several types of digital biomarkers for the management of AD. A more detailed overview of how features captured by these devices could be used for AD management is given in this section.

5.2.1.1 Devices based on cognitive tests

Current early detection trials use expensive scans and extensive methodologies that cannot accurately test asymptomatic patients with AD. Furthermore, it is often difficult to evaluate the designated treatment's effectiveness in pre-clinical AD or mild cognitive impairment cases due to the difficulty in assessing cognition. Doraiswamy *et al* (2018) proposed that automated mobile cognitive tests may lead to automated and precise scoring, coupled with features such as multiple test versions and individualized scaling. Additionally, the amalgamation of these factors with an element of gaming could create a technology that would lower the anxiety associated with the testing mechanisms.

Rentz *et al* (2016) developed an iPad version of self-testing named C3-PAD (Computerized Cognitive Composite for Preclinical Alzheimer's Disease) and tested it in 49 subjects with normal cognitive functioning. Results indicated a significant association ($r^2 = 0.508$, $p < 0.0001$) between the outcome of the C3-PAD assessment conducted in a clinic and at home. The moderate correlation in results concerning

C3-PAD evaluations and standardized tests, which covered similar cognitive domains, indicated more robust results for the C3-PAD than the standardized tests. These results demonstrate how certain cognitive functioning aspects can be assessed more reliably at home, serving as a viable addition for older patients with impaired mobility. It further highlights the need to modify the pre-existing, standardized tests.

5.2.1.2 Devices based on motor functioning

Signs of motor impairment are usually noticeable a decade before cognitive impairment becomes apparent in patients with AD (Albers *et al* 2015). Deviance in fine motor skills such as finger tapping, typing, and drawing can indicate mild cognitive impairment linked with AD. The Mini-Mental State Examination (MMSE), a 30-point questionnaire, is used widely to assess cognitive impairment and screening for dementia. An MMSE score greater than 24 is considered normal, while 20–24 is regarded as mild dementia. Rabinowitz and Lavner (2014) assessed the link between finger tapping and cognitive impairment among 170 elderly subjects by making them undergo MMSE, forward digit span test, and 15 seconds of finger tapping. They observed an increased length and variability in finger tapping in subjects with mild cognitive impairment or dementia compared to other subjects who showed no prevalence of impairment. This result suggests a possible correlation between finger tapping, attention, and short-term memory. The link between tap numbers and gait features indicates the possibility of a connection between the ability to perform motor tasks and functioning capacity.

How older patients interact with technology can indicate variation from normal cognitive functioning. Computer behaviors in individuals with cognitive impairment are substantially different in comparison to cognitively healthy individuals. These behaviors include frequent pauses while typing, a slower typing rate, and a more significant proportion of mouse clicks. These behaviors are linked to memory-related performance on cognitive functioning assessments. For example, 128.48 ± 35.03 keystrokes per minute were observed in subjects with mild cognitive impairment and 63.65 ± 32.64 keystrokes per minute in healthy individuals (Stringer *et al* 2018). Thus, monitoring the pattern of computer behaviors using semi-directed computer tasks among older patients could detect AD. Such software could be developed and incorporated into devices. One could also study the digital stylus's motion through tracing tests by measuring deviation from the shape to assess cognitive function.

In addition to fine motor skills, gross motor skills also can be an indicator of mild cognitive impairment. Cognition and gait functioning have a common anatomical control, mainly the prefrontal cortices. Changes in gait (e.g. slowed gait) could predict cognitive decline (Cohen *et al* 2016). Smartphone applications, wristbands, watches, patches, and rings could be used to measure stride length and step count accurately. In addition to this data, additional information could be collected by incorporating sensors for geo-positioning. After carefully curated feature engineering, data from these devices would be invaluable in developing predictive models to detect early symptoms of AD. For example, in a study by Jian (2020), predictive models using support vector machine regression and linear regression techniques were developed and tested using features from the finger tapping (number of taps, average tapping interval, tapping frequency) and the Time Up and Go (TUG) test (gait speed, pace, speed variance, acceleration range). They built predictive models to predict the TUG score from these features. The support vector machine regression model demonstrated a mean average error of only 1.218 using features from both tests. The results suggested a strong correlation between finger tapping and cognition and the link between the risk of falling and gait characteristics obtained from the TUG score.

5.2.1.3 Devices based on oculomotor functioning

Pupillary reflexes and movement of the eyes have been relied upon for research of neurological conditions for decades. Assessing these aspects assists in probing aspects of the temporal lobe, brain dopamine activity, and cholinergic pathways of neurons and identification of neuropathological changes. The movement of the eye generates a pattern while performing tasks such as reading. Every eye movement results in a fixation point which permits the brain to assess the inflow of information and program the rapid eye movement between fixation points (saccade). However, in patients with AD, these eye movements tend to be abnormal or disturbed. A study performed on 40 participants (20 controls, 20 patients with AD) concluded that even in the early stages of AD, eye movement is altered while reading (Fernández *et al* 2013). AD patients exhibit a reduction in the number of words per fixation, an increase in the frequency and duration of fixations, and an increase in the number of words missed. Another study conducted on 70 participants (35 controls, 35 patients with AD) noted that preceding words' predictability is highly predictable.

Thus, monitoring the eye movement and using the eye movement-related features in predictive models can help in the early diagnosis of AD. For example, Biondi *et al* (2017) trained a deep learning model called the denoising sparse-autoencoders to identify patients with AD using eye-tracking features acquired when the subjects read validated sentences and proverbs. They reported an AD detection accuracy of 89.78%.

5.2.1.4 Devices based on sleep and circadian rhythm

Patients with AD are associated with having disturbed sleep and circadian rhythm patterns, which may be prevalent as agitation in the night. Patients with AD have also been shown to exhibit a reduced rapid eye movement sleep (REM). Since this stage helps retain long-term memories, any REM stage disturbances contribute to memory issues (Bliwise 2004). An observation study by Lim *et al* (2013) showed that fragmented sleep is linked with AD and the rate of decline in cognition. Pervasive devices such as the Oura ring, a multisensor sleep-tracker, could potentially be used to track sleep patterns in a home setting. de Zambotti *et al* (2019) validated the ŌURA ring against polysomnography (PSG), the gold standard in sleep studies to record brain waves, blood oxygen content, heart rate, and movements during sleep. The Oura ring showed a sensitivity of 96% in sleep detection and a specificity of 48% in detecting awake states. It was also able to detect various sleep stages and total sleep time. Many sleep variables were quantified the same by both the ring and PSG, making it a device that can potentially monitor sleep in a home setting.

Wrist-worn devices that combine optical and inertial sensors to estimate the user's breathing rate, pulse, and movement could also be used to compute if the user is in REM, non-REM, or awake states. Reveney *et al* (2017) assessed a wristband's functioning by comparing the results to those produced by polysomnography. The method achieved a sensitivity of 89.2%, a specificity of 77.9% for the REM stage, and a sensitivity of 83.4%, a specificity of 84.9% for the non-REM stage. Similarly, most wrist-worn devices and app-based monitors show comparable reliability when compared to polysomnography.

Again, data from these wireless sensor-based devices can also be used in predictive modeling. An example is a study published by Boe *et al* (2019). They proposed a wireless multimodal sensor system that measures hand acceleration, electrocardiography, and distal skin temperature. They used the measured parameters to build and test a bagging classifier to detect wake and sleep stages. The proposed methodology outperformed a commonly used sleep monitoring wrist-based device called the Actiwatch.

5.3 Parkinson's disease

PD is the second-most prevalent neurodegenerative disorder affecting an individual's ability to perform motor functions such as walking and maintaining body balance. PD occurs due to damage or death of neurons in the brain's motor regions, and PD-linked neurodegeneration occurs years before the symptoms emerge. Being a chronic disease, PD symptoms (such as rigidity, resting tremor, and bradykinesia) worsen with time. In addition to motor function decline, people with Parkinson's disease (PwPD) could also exhibit changes in behavior, depression, and a drop in recall ability (memory function). Several symptoms of PD can be attributed to the loss of neurons responsible for dopamine production. Dopamine is the brain's chemical messenger that could lead to abnormal brain activity when present at low levels. PD risk factors include the individual's age, genetic component, gender, and exposure to toxic elements such as pesticides.

Often, the initial symptoms are subtle and could be perceived as typical aging characteristics, thereby delaying diagnosis until a more progressed PD stage is reached. The patient's or associated caregiver's recollection of the severity of motor and non-motor symptoms is relied upon heavily to recommend treatment. However, this could be inaccurate due to the lack of clinical knowledge (inability to differentiate tumors from dyskinesia) or the failure to record the frequency of symptoms and their progression. Freezing of gait (FOG), stooped posture, shuffling steps, festination, and falling are the specific disorders of the gait observed in PwPD (Chen *et al* 2013). Several solutions have been used to study these gait disorders—motion-sensing camera-based technologies, specialized treadmills with various sensors to quantify tremors, pressure measuring devices, EMG systems, etc. The key disadvantage of these clinic-based devices is that they are limited-time studies.

To more accurately diagnose and monitor the progression of the disease, continuous monitoring is required. In this regard, wearable pervasive devices embedded with predictive algorithms can be beneficial for diagnosing and monitoring the disease and managing or reducing the symptoms. Rovini *et al* (2017) provide a detailed review on wearables used for early detection, tremor and body motion analyses, the study of motor fluctuations, and remote monitoring. The next section provides a brief overview of several types of pervasive devices for PD diagnosis and management.

5.3.1 Pervasive devices for PD

5.3.1.1 Devices for early detection

Currently, the diagnosis of PD is based on the physician's assessment of the patient's performance of certain specific tasks in the clinic. The physicians then assign scores based on scales such as the Unified Parkinson Disease Rating Scale (UPDRS) or Hoehn and Yahr scale. Because of inter-observer variability, the diagnosis error is generally around 40% (Channa *et al* 2020). Pervasive devices could provide for a more objective assessment of the diagnosis and stage of the disease. Lan and Shih (2014)proposed a smartphone-based early warning tool that utilized the accelerometer sensor embedded in the smartphones. The analysis of the sensor data such as step length and step frequency presented insights into the patient's gait characteristics, which were then used for early diagnosis. Perumal and Sankar (2016) investigated the use of gait features such as step length, stride time, swing time, etc, and heel force, below toe force, etc, detected from force sensors placed below the toe, and tremor features such as amplitude, power, and frequencies of the tremor signals, and used them to train machine learning algorithms to detect the presence of PD. Margiotta *et al* (2017) developed a wearable wireless system composed of inertial measurement units (IMUs) placed on the patient leg to capture and analyze gait characteristics for the early detection of AD and PD.

5.3.1.2 Devices for the management of symptoms and treatment

5.3.1.2.1 Tremor

Tremor is often the earliest symptom of PD. It is an involuntary movement with a 4–6 Hz frequency (Elias and Shah 2014). Tremor analysis helps detect different types of tremors using the amplitude and frequency of the tremor. The type of treatment depends on the correct diagnosis of tremors. Therefore devices that can capture and analyze parameters such as amplitude, frequency, and fluctuations are essential. One tremor monitoring device is the Kinesia 360 that uses wearable sensors, and the Kinesia 360 mobile app to continuously and objectively measure tremors using accelerometers and gyroscopes embedded in a small sensor device worn on a finger (Giuffrida *et al* 2009). Other sensor systems generally used for tremor analysis are the goniometer, optical motion capture system, and IMUs. These devices help identify the stages and types of PD (Rigas *et al* 2012, Daneault *et al* 2013).

5.3.1.2.2 Gait

FoG is defined as a 'brief episodic absence or marked reduction of forward progression of the feet despite the intention to walk' (Giladi *et al* 2013). The complex pathophysiology of FoG and the variety in cases result in the lack of consensus on a common treatment option among approaches such as surgery, physiotherapy, pharmacology, and occupational therapy (Nonnekes *et al* 2015). Conventionally, the treatment options for PD include the prescription of medications (dopamine precursors, dopamine agonists, monoamine oxidase inhibitors among others). However, these 'antiparkinsonian' medications can cause unwanted psychomotor and autonomic complications (Manson *et al* 2012). On the other hand, invasive surgical procedures are expensive and may not be feasible for some PwPD. Therefore, there is a need for improved treatment options for FoG. One such treatment option used to manage the symptoms is cueing techniques (Sweeney *et al* 2019).

Cueing is a popular non-invasive technique that uses external stimuli to use spatial or temporal information to enable or improve movement (Nieuwboer *et al* 2007). It is achieved by increasing the walking speed, stride length, and cadence (the rate at which a person walks measured by steps taken per minute). Auditory, visual, and somatosensory cueing are the main cueing techniques. Combining two techniques such as the auditory and visual stimulus, to obtain an audio-visual system (BodyBeat Pulsing by McCandless *et al* (2016)) can also be implemented to provide a rounded approach to cueing. Several theories attempt to reason the effectiveness of cueing as a strategy to alleviate FoG. According to studies, the task of moving could be influenced by the compensatory role of cueing for the defective internal rhythm generator of the basal ganglia (Azulay *et al* 1999).

5.3.1.2.3 Visual cueing

In the simplest form of visual cueing, Dietz *et al* (1990) demonstrated visual cues to diminish FoG episodes by marking parallel lines on floors for the patients to walk on. A gap of 305 mm between the parallel lines provided spatial information to the PwPD. It is unclear whether the correlation between the parallel lines and reduction in FoG episodes is due to alteration of step length, conscious attention towards the task, or enhanced proprioception in the inferior aspect of the body.

Table 5.2 lists a few types of visual cueing devices developed over the last decade. Espay *et al* (2010) developed a home-based visual-auditory walker which could

Table 5.2. Summary of visual cueing devices for PD.

Author	Device
Espay *et al* (2010)	Visual-auditory walker
Bryant *et al* (2010)	Laser walking cane
Donovan *et al* (2011), Bunting-Perry *et al* (2013), McCandless *et al* (2016)	U-step walking cane and U-step walking stabilizer
Buated *et al* (2012)	LaserCane device
Tang *et al* (2017)	Chest-worn laser device
Barthel *et al* (2018)	Laser shoes
Ahn *et al* (2017)	Smart gait-aid system (application for smart glasses)

continuously project a virtual checkerboard-tiled floor that responded to the user's movements. The tiles acted as targets for the user to step on. The system consisted of a sensor unit and a head-mounted micro-display with earphones that provided auditory feedback. A small study indicated that the system could improve could reduce FoG episodes. Bryant *et al* (2010) proposed a walking cane that projected the horizontal lines using lasers and determined in a pilot study that the green laser line could positively reduce FoG episodes.

Donovan *et al* (2011), Bunting-Perry *et al* (2013), and McCandless *et al* (2016) evaluated the U-Step walking cane and U-Step walking stabilizer, which were commercially available. The U-Step walking cane had a weight-activated switch that projected a red laser line when the user put force on the cane, and the U-Step walking stabilizer was a four-wheeled rolling walker with a red laser attached at ankle level. Although these types of visual cueing devices reduced FoG episodes, these studies concluded that they should be simple to use with single-tasking instead of multi-tasking for the devices to be useful. Also, appropriate training should be provided to the users. Buated *et al* (2012) and Tang *et al* (2017) also developed and tested similar cueing devices. Most of these visual cueing devices were not easy to use in daily life. Thus in 2018, Barthel *et al* (2018) designed laser shoes that would project the cues and were less cumbersome for the users to wear. They also reported positive effects on the frequency of FoG episodes and more user acceptance.

In a study by Ahn *et al* (2017), the team developed a real-time FoG detection algorithm based on cadence and length of stride customized to be used in Epson's Moverio BT-200 smart glasses. Even though the FoG detection algorithm itself had a sensitivity of 97% and specificity of 88%, the system was limited by computational and battery capacities.

5.3.1.2.4 Auditory cueing

Around 2010, auditory cueing was generally done using devices called metronomes. Metronomes emit an audible sound at regular intervals, which in this case could be used to indicate the step duration to the PwPD. The limitation is that metronomes continuously emitted cues irrespective of the existence of an FoG episode. This limitation led to the development of systems that only provided auditory cues in the presence of FoG. FoG detection algorithms using signal processing techniques such as the fast Fourier transform (applied on the data from acceleration sensors) were embedded in them (Bächlin *et al* 2010). The limitation of these systems was the varied sensitivity and specificity of FoG detection among participants.

Mazilu *et al* (2015) developed a home-based system called the GaitAssist system to provide motor training and gait assistance at home. The GaitAssist system comprises three components, a smartphone and two sensors attached to the user's ankles. They implemented an app in the smartphone that could perform frequency analysis and machine learning-based detection of FoG from the signals received from the ankle sensors. Once an FoG episode was detected, the user was alerted by a rhythmic ticking sound via headphones. The FoG detection algorithm had a good sensitivity of 97.1%. Another advantage of the system was that the app had an automatic capability to adapt the cueing tempo to the user's cadence in real-time. A small trial was conducted to evaluate the device on five participants for three days. Four participants displayed a reduction in the duration and frequency of FoG episodes. The limited sample size does not provide an accurate depiction of the efficacy of the device. Additionally, a subjective questionnaire was provided to another set of nine participants, who used the device for approximately three hours a day for one week. Users reported two minor shortcomings—difficulty attaching the ankle sensors independently and participant preference in using smartphone speakers instead of headphones.

5.3.1.2.5 Audio-visual cueing

In 2016 Google and Microsoft introduced smart glasses, wearable devices that look like traditional glasses but contain features similar to smartphones. They could be equipped with accelerometers, Wi-Fi connections, GPS, and audio-visual output features. The structuring of the glasses poses the possibility of three-dimensional cueing. Additional technological features include voice commands and gestures that make smart glasses hands-free, making them well-suited for PD patients.

Zhao *et al* (2016) created an application for Google Glass, which can provide three types of audio-visual cues: metronome, flashing light (LED), or optic flow. The frequency range was fifty to one hundred and fifty cues per minute. The metronome was a purely auditory cue that provided rhythmic beats. The optic flow was a visual stimulus that provided cues by generating vertical lines on both sides of the screen, propagating at a determined speed (measured in lines per minute).

Twelve PwPD who had a history of a minimum of two FoG episodes per day were selected. Participants with major cognitive impairment were excluded from the study. Only those who could walk a distance of twenty meters on a flat surface without needing walking assistance were included. Cueing was carried out when the participants were at the end-of-dose stage (3.04+/−0.84 h post-consumption of last medication).

The cadence was calculated over a walking period of ten seconds. Participants were permitted to alter the device's cueing frequency to their preferred walking speed and were then directed to match each step to a beat of the metronome. For testing LED as a method, the participants were asked to take a step when the screen flashed on and another step when it flashed off. Lastly, for the optic flow, participants were asked to take a step with the leg corresponding to the side of the glass where the moving bar appeared. Thus, a step with the left foot would be taken when the bar would appear on the left side of the screen, and a step with the right leg would be taken when the bar appeared on the right side of the screen. The cueing frequency was uniform for auditory and visual cueing (average = 106.1+/−11.5; range = 80–124 steps per minute).

The results indicated that auditory cueing had a more significant impact than visual cueing. Based on a questionnaire administered post-trial, 75% (n = 9) of patients agreed that they would be willing to incorporate the use of the Google Glasses in their routine. Additionally, seven and nine participants reported that operating the device was easy or very easy. The instructions provided on the screen to read were clear or very clear, respectively. A unique feature, particularly liked by a participant, was the bone-conduction speaker that made the metronome less audible to those in the participant's surroundings. This feature enhances the rehabilitation of the patients in social settings and makes it a feasible device. However, the device cannot detect FoG episodes, and the small sample size makes the data inconclusive. Moreover, it was observed that the cues were produced at a fixed rate that did not change with the user's motion speed causing the user to probably experience dizziness and loss of balance. Regardless, it demonstrates the scope of using smart glasses to assist in pervasive cueing strategies.

5.3.1.2.6 ***Somatosensory cueing***

In somatosensory-based cueing, the devices provide rhythmic vibratory stimulations to the user. As early as 1997, Enzensberger *et al* (1997) proposed using somatosensory stimulus in the form of repeated shoulder tapping to set a rhythm among PwPD. The tapping provides temporal information such as step duration. Rosenthal *et al* (2018) built a cueing device called the cueStim, a voltage-controlled two-channel electric simulator to reduce the FoG symptoms associated with PD. The device was worn around the waist and delivered fixed rhythmic sensory stimulation to the patient via electrodes placed on the muscles most affected by PD. The device was tested on nine patients with a mean age of 73.33 years, and it was observed that a fixed rhythmic sES cueing strategy reduced the time taken to complete a walking task and the frequency of FoG episodes. On reviewing the questionnaires administered post the assessment, it was noted that the patients' average rating of comfort while using cueStim was 'comfortable', and the pain rating was 'no pain'. However, the small sample size and variation in FoG episodes among participants make it difficult to generalize and assess the effectiveness of cueStim.

5.3.1.2.7 ***Combination of symptoms***

Personal KinetiGraph® system (PKG) by Global Kinetics Corporation

The Personal KinetiGraph Movement Recording System aims to constantly and objectively quantify the disabling symptoms, including tremors, bradykinesia (BKS), and dyskinesia (DKS), associated with movement disorders such as PD. The system comprises a PKG watch, an interactive data logger resembling a standard wristwatch that is worn on the wrist corresponding to the side of the body that has been more affected by the symptoms (Santiago *et al* 2019). The watch is equipped with an accelerometer, a battery, and memory to store data, an optional reminder for medication consumption, a log of when the medications are taken, and a capacitive sensor that detects the removal of the watch from the wrist. The PKG data logger can be configured easily for a PwPD by clinical staff through a tablet computer program. The patient's PKG watch must be worn for seven days, following which the data of PwPD's recorded daily movements are downloaded for processing. The raw data collected are then analyzed using an algorithm provided by the manufacturer to generate a printable version (PKG) of the tracked movement which incorporates summary scores for DKS, BKS, fluctuations, tremors, immobility, daytime movement, sleep, daytime somnolence, and the duration wherein the device was not worn. These details are then reviewed by a clinician who decides on the next clinical steps for the patient. The target group for the PKG system is individuals who are forty-six to eighty-three years old. The PKG system has received regulatory clearance through FDA 510 (k) in the USA, CE mark in Europe, and is TGA Class IIa ARTG 236583 listed in Australia.

The PKG device has been validated in several studies. In one recent study by Santiago *et al* (2019), 112 physician surveys were analyzed, and it was observed that at least a third of the time, the physicians used PKG data to improve the treatments for the patient. In another study by Joshi *et al* (2019), 63 PD subjects were enrolled to study the benefit of using a PKG device in a clinical setting. The most critical benefit observed was that the PKG could pick up 50% of BKS and 33% of DKS symptoms that failed to be reported by patients. Other benefits were improved patient dialogue, improved assessment of treatment efficacy, and motor capabilities.

Insole based devices

Insoles in shoes are much cheaper wearable devices and are relatively unobtrusive compared to other devices. These insoles can use inertial sensors such as accelerometer, gyroscope, piezoelectric, and force sensors (Channa *et al* 2020). Studies have been conducted using insoles to capture features such as stride time, step length, variations in plantar pressure, sway, rotational angles of feet, etc, and the features have been used to train several machine learning algorithms to classify PD stages and also in symptom monitoring (Ngueleu *et al* 2019). Smart insoles have shown to be 75% to 100% accurate and could potentially be a good candidate for the objective assessment of PD.

5.3.2 Predictive algorithms

As noted before, subjective clinical scales and patient-based diaries have limited use in PD disease detection and management because they do not provide continuous real-time data about the patient's disability. The use of wearable continuous monitoring devices results in the accumulation of large volumes of data. These datasets could be used to build predictive algorithms using machine learning and deep learning techniques. For example, Ramdhani *et al* (2018) describe how several machine learning algorithms such as random forests, decision trees, logistic regression, support vector machine, hidden Markov models, naive Bayes, and neural networks have been implemented successfully in PD management. They review algorithms that use data from inertial sensors that are constructed from accelerometers and gyroscopes. In another study Lonini *et al* (2018) captured movement data using six flexible wearable sensors. They used them to train convolutional neural networks and statistical ensembles to classify segments of signals to a type of symptom.

5.4 Epilepsy

Epilepsy is a neurological condition affecting nearly 50 million people across the globe (WHO 2019). It is marked by an abnormality in brain activity that results in seizures. Although antiepileptic drugs can control the seizures associated with epilepsy, more than 30% of individuals with epilepsy present with drug-resistant cases of seizures (Sheng *et al* 2018). Seizures affect the brain, which can be associated with symptoms such as uncontrolled jerks in the arms and legs, temporary confusion, loss of consciousness, and staring spells. The current treatment and management protocols for epilepsy depend on the patients' recollection ability regarding the frequency of seizures and related symptoms in the form of self-reporting. Patients or their families may not notice a seizure or may believe seizure-like symptoms are a seizure. This unreliable reporting could lead to an inaccurate capture of patient history and incorrect treatment recommendations.

Support regarding the detection of epilepsy and subsequent management is vital to prevent its progression and accidental injuries or sudden unexpected death in epilepsy (SUDEP). Pervasive devices developed for patients with epilepsy can detect epileptic seizures, alert for the requirement of medical assistance for individuals, provide patients with details of the treatment process, and provide assistance regarding safe self-management of the disorder remotely. Seizures are generally classified into generalized seizures (abnormal activity involving the entire brain) and focal seizures (abnormal activity in a specific area of the brain). Generalized seizures, which are the primary focus of most devices for the detection of epilepsy, can be subdivided into the following:

1. *Tonic seizures* (TS): causes stiffening of muscles, resulting in falls.
2. *Clonic seizures* (CS): causes rhythmic jerks in muscle movements.
3. *Tonic-clonic seizures* (TCS): causes body stiffening, loss of consciousness or tongue-biting.
4. *Atonic seizures* (AS): causes loss of muscular control, resulting in falls.
5. *Myoclonic seizures* (MS): observed as sudden jerks or twitches in arms or legs.

The compliance of pervasive devices among patients can be attributed to their appearance, comfort, compactness, low intrusiveness, the privacy of data collected, and proper functioning. Most technologies that report seizures indicate the need for better devices that can distinguish between the types of seizures and accurately ascertain the frequency of seizure occurrence. Furthermore, false-positive detection of seizures while carrying out routine activities like standing and sitting in some of these devices makes them less compatible and reliable than conventional self-reporting methods. Furthermore, it could impose restrictions on the patient to reduce the false alarm rate (FAR). Additionally, devices that can predict seizures are not available. High detection sensitivity and FAR are vital factors that indicate the performance of devices.

5.4.1 Typical framework

Currently available devices are primarily focusing on incorporating non-EEG signals. Devices used to detect and alert for seizures usually contain sensors such as accelerometer (ACC), gyroscope, electrodermal activity (EDA), temperature, and photoplethysmograph (PPG) sensors. An ACC monitors the movement and provides information regarding the orientation of a device and, therefore, the patient. The gyroscope provides an additional dimension by indicating the rotational movement (angular velocity). A gyroscope is a sensor that varies depending on the phone's orientation based on the movement of the phone made by the user. EDA measures changes in sweat gland permeability by obtaining differences in electrical potentials in varying parts of the skin. The PPG optical measurement technique is an inexpensive method of monitoring heart rate. It is non-invasive and employs a photodetector and a light source at the skin surface to measure the volumetric variations of blood circulation (Castaneda *et al* 2018). The following section describes three more recently developed and validated commercial devices used for epilepsy management, giving the reader an idea of how pervasive devices are used for epilepsy care.

5.4.2 Pervasive devices for epilepsy

5.4.2.1 Epi-Care Free by Danish Care Technology, Denmark

Epi-Care is a device created to be worn on the wrist that detects seizures using an accelerometer that detects the vibrations and a gyroscope that distinguishes between daily activities and seizures to ensure only seizures are detected (Beniczky *et al* 2013). Epi-Care technology consists of the Epi-Care Free and the Epi-Care Mobile component. Epi-Care Free is a wristband that detects tonic-clonic seizures using in-built sensors and has a wirelessly connected portable base unit. This base unit can receive an alert from the wristband during a seizure within a range of 20 m and consists of a journal that tracks the epileptic events by date and time. However, the range can be expanded by the Epi-Care Mobile device, which sends the alarm further to a family member or healthcare provider present beyond the 20 m radius. Epi-Care is CE certified and is available commercially in the United Kingdom, Ireland, and mainland Europe.

5.4.2.1.1 Clinical trials

The Epi-Care Free system was clinically tested at three centers: Epilepsy Center Bethel, Bielefeld, the Danish Epilepsy Hospital, Filadelfia, and Rigshospitalet University Hospital. The trial included 73 participants who wore the Epi-Care wristband, coupled with real-time video-EEG monitoring, to compare the results and determine the accuracy. The three-day trial resulted in the detection of 35 seizures out of 38, i.e. 89.7% of total tonic-clonic seizures and an FAR of 0.2 seizures per day (Beniczky *et al* 2013). The blinded nature of the study prevented the seizure-alerting feature (Epi-Mobile) from being assessed.

A field study conducted by Meritam *et al* (2018) incorporated both the Epi-Care Free and Epi-Care Mobile features in their trial. In the phase 4 trial, 112 participants were provided questionnaires. These 112 participants were registered Epi-Care users. The questionnaire consisted of ten questions gathering data about the patient background and 15 questions about the ease of operating and functioning of the Epi-Care technology. 63% (n = 71) of patients/their caregivers completed the questionnaires and the structured interview themselves. The patients' device usage ranged from 24 days to 6 years, with a median of 15 months. The accuracy was calculated by comparing the device count to the description of seizure instances provided by the patients/caregivers compared to video-EEG monitoring used in the clinical trial. The median sensitivity observed was 90%, with an FAR of 0.1 seizures per day. Seven patients discontinued the use of the device for the following reasons: (a) a high FAR rate ($n = 4$), (b) inability to operate the device ($n = 2$), and (c) a damaged device that was not replaced ($n = 1$). 89% ($n = 63$) of patients experienced no side effects of using the device. However, in four cases, a rash or skin irritation occurred from the wristband, and in two cases the device was found to have disturbed the functioning of other home appliances.

5.4.2.1.2 Evaluation

Positives:

- The wristband needs to be charged for merely 1 h day^{-1}.
- Does not require constant observation.
- Easy to use.
- Routine activities such as eating are not recorded by the device, thereby preventing unrequired alerts or seizure counts.
- The alerting feature alerts the caregiver to assist in a seizure.
- Fewer injuries to the patient (as suggested by 40% of participants in the phase 4 trial).
- The highest seizure count detection rate among accelerometer-based epilepsy devices.

Negatives:

- Can only be used to detect tonic-clonic seizures.
- The inability to operate the device due to the lack of technological knowledge or inability to set the device components effectively. Additionally, the receptivity of older adults towards learning how to use technology could prove to be a significant factor.
- Not water-resistant.
- Possible rash or skin irritation.
- Epi-Care Free does not function normally if any assistance is provided during the seizure (e.g. nurses holding the patient's hand for support where the device is placed).

The lack of constant observation improves the quality of life of patients with epilepsy by providing them more freedom, mobility, and peace of mind. The alerting feature and the in-built journal prevent the loss of count of seizure frequency, a vital component to diagnose the severity and management of epilepsy. 40% of participants observed a reduction in injuries while using Epi-Care. According to Meritam *et al* (2018), the patients and caregivers were satisfied with the device, which is influential in determining patients' compliance.

5.4.2.2 NightWatch by LivAssured BV, the Netherlands

NightWatch is an armband device designed to be used when the patient is asleep during the night when seizure occurrences are likely to be unanticipated or go unaccounted for. The upper-arm position was preferred over the wrist due to enhanced signal quality and fewer movement artifacts (Arends *et al* 2018). The band is coupled with a small box that contains the PPG sensor and the three-axis accelerometer sensor, which monitor the change in heart rate (HR) and the change in motion, respectively. NightWatch uses these sensors to detect seizures with major motor characteristics through an in-built algorithm. NightWatch detects various seizure types (tonic, tonic-clonic, hyper motor, clustered myoclonic). The alarm activates when the person is lying down. Suppose the HR threshold exceeds a threshold (i.e. the patient has tachycardia or when the motion value remains above the threshold for a minimum of 15 seconds). The potentially severe seizure triggers a warning signal transmitted to the caregiver through a wireless signal (Digital Enhanced Cordless Telecommunications, DECT) to the armband's base station. NightWatch has received CE approval as a Class I medical device, which is commercially available in the European countries that accept CE certification for medical devices (Bruno *et al* 2020).

5.4.2.2.1 Clinical trials

Arends *et al* (2018) carried out a prospective, in-home study to detect nocturnal seizures among 34 admitted participants with more than one seizure a month by asking them to use the NightWatch on the upper arm for up to three months. Twenty-eight participants completed the study and 809 major seizures were detected, and 2040 alarms were received. The reliability of the device was tested by an inter-observer agreement, and the performance of the sensor was compared to a bed sensor (EmFit; a thin sheet placed under a mattress). A median sensitivity of 86%, an FAR rate of 0.03 seizures per night, and a positive predictive value of 49% were observed. However, the bed sensor produced a lower sensitivity in comparison to the NightWatch. 368 of 1402 (26%) false-positive alarms were minor seizures. Some patients reported skin irritation, and the irritation was resolved by removing the paint on the surface of the device and ensuring that the NightWatch was not tight around the arm.

5.4.2.2.2 Evaluation

Positives:

- The accelerometer prevents alarms from going off during routine activities like sitting or standing.
- Easy to use.
- Positioning on the arm instead of the wrist promotes optimal body contact to determine HR.
- A greater signal quality and lower error rates in comparison to smartwatches.
- No smartphone is required.
- The DECT wireless connection provides a range of 50 m (greater than Bluetooth, which has a range of 10 m).
- It can detect more than one type of seizure.

Negatives:
- Seizures other than the tonic-clonic kind are less well detected (Arends *et al* 2018).
- Minor seizures are sometimes counted as false positives.
- Not water-resistant.
- There is a lack of video-EEG usage to compare the results of the trial.
- Daytime use may count activities as seizures, thereby contributing to false positives.
- False positives are more likely in patients with high resting HR values or children who tend to be highly excitable (Fleming *et al* 2011).

According to the caregivers in the trial, the sensor alerts enabled timely help in urgent situations. Furthermore, a majority of caregivers believed that the device provided them with more freedom. The sensor functioning issues observed in the trial were temporary or resolved. Minor seizures were considered false alarms in the study, and alarms occurred only during more severe seizures that demanded attention from the caregiver. Since minor seizures are valuable to assess the frequency and pattern of seizures in a patient, a reformed algorithm can be designed to distinguish between both the seizures based on severity and non-seizure movements. The multimodal sensor may potentially furnish information relevant to determining the urgency of seizures since increased HR height and duration have been found to correspond with an estimate of speed (Van Andel *et al* 2015). Signals from muscles captured by a wearable electromyogram (EMG) device have also been shown to help detect seizures (Beniczky *et al* 2018).

5.4.2.3 Embrace, the wrist sensor by Empatica

Embrace is a lightweight wearable device that provides an automated platform to count seizure numbers accurately, provide alerts to indicate the need for intervention, and offer constant ambulatory monitoring. Ambulatory monitors are compact and mobile electrocardiograph devices that assess the rhythmicity of the heart over time. The Embrace and E4 wristbands developed by Empatica are the first of their kind to be accessible commercially (Regalia *et al* 2019). These multimodal wristbands function by detecting the physiological features of a generalized tonic-clonic seizure. Both the devices have accelerometers to identify motion (ACC) and electrodermal activity (EDA). Embrace is a prescribed device that is also equipped with an angular velocity sensor (gyroscope). The wristband alerts caregivers through the 'Alert App', a smartphone application via Bluetooth in the form of a phone call/text message (Bruno *et al* 2020). Subsequently, the system stores the data acquired from the ACC, EDA, and temperature sensors to be reviewed by healthcare professionals. The 'Mate App' helps in the daily management of epilepsy by enabling seizure tracking, medications, daily activity, and rest tracking. Embrace has received clearance from the United States Food and Drug Administration (FDA) as a Class II device for adults and children above six years old in 2018 and 2019, respectively (Regalia *et al* 2019). Both Embrace and E4 have received the CE certification (Rukasha *et al* 2020). Embrace also uses a machine learning-based detection algorithm that uses the ACC and EDA data.

5.4.2.3.1 Clinical trials

A single-center study assessed the prototype measuring EDA and ACC among 80 patients in the epilepsy monitoring unit (EMU). The 127 day (4213 h) study employing a semi-patient specific algorithm resulted in a sensitivity of 94% (15 out 16 seizures detected) and an FAR of 0.74 seizures per day (Poh *et al* 2010). This study was not submitted for the FDA clearance of Embrace.

A clinical trial of Embrace was carried out on 114 patients with epilepsy who were admitted to the EMU. The patients included 80 children and 61 adults ranging from ages 6 to 21 and 22 to 63 years. 78% of EMU patients (n = 110) experienced no seizures, while the others (n = 31) experienced 54 generalized tonic-clonic seizures in all. The data were collected for 409 days (9806 h) which approximates 49.2 h of data per patient. The results indicated a sensitivity of 98% (53 out of 54 seizures detected) and an overall FAR of 0.94 seizures per day (Bruno *et al* 2020). The study was submitted to the FDA by Empatica.

5.4.2.3.2 Evaluation

Positives:

- Water-resistant to up to one foot.
- Lightweight (13 grams).
- Four sensors (EDA, temperature, ACC, gyroscope).
- Available worldwide.
- Easy to use.
- No ionizing radiations.
- No localized heat.

Negatives:

- Only detects generalized tonic-clonic seizures.
- Not foolproof (false positives and false negatives occur).

5.5 What is next?

This chapter provided an overview of pervasive wearable devices for managing three common neurological disorders, namely Alzheimer's disease, Parkinson's Disease, and epilepsy. The chapter aimed to present a general idea of the nature of the disease and the features/symptoms captured and monitored by these wearable devices to either diagnose the condition early or manage the symptoms effectively. Improving the quality of life is key to managing these disorders successfully, and continuous monitoring is not a feature of many existing clinic-based technologies. The chapter also reviewed the advantages and limitations of several pervasive devices. It also outlined how predictive analytics can play a role in harnessing the power of data from these wearable devices to predict, diagnose, and manage the disorders more objectively. This review is by no means comprehensive, and interested readers are directed to the references to learn more about a specific topic.

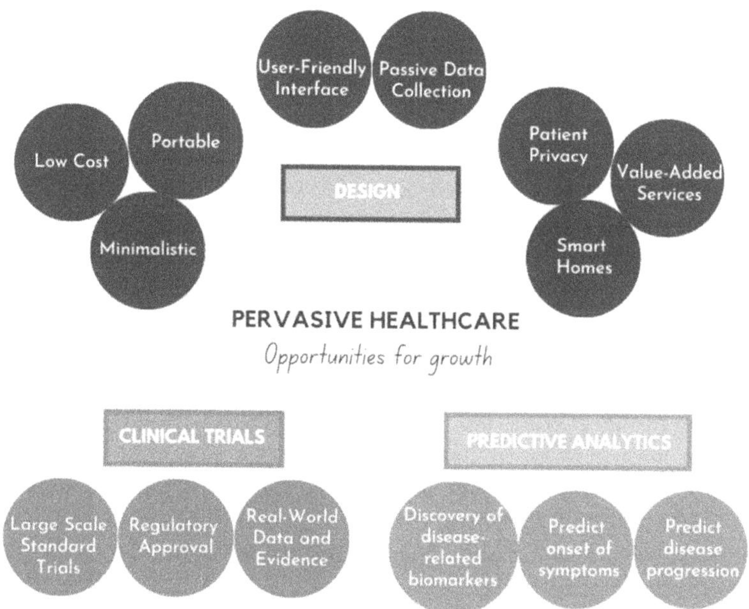

Figure 5.1. Outline of opportunities for growth in pervasive healthcare.

Even though wearable devices may not be very suitable for patients with severe cases of these diseases, such as severe motor and cognitive impairments, these devices can detect these diseases early and help improve the quality of life of patients. During the global COVID-19 outbreak, patients have found it challenging to go to hospitals to obtain the medical care they need. Physicians have also been unable to monitor their patients closely. Wearables and telemedicine have played a significant role in this scenario by enabling remote monitoring.

5.5.1 Challenges and advancements necessary in the technology space

As more solutions are developed in this space, some key advancements are necessary for the technology and application areas. Below is an outline of opportunities for growth from a technological standpoint, and a visual summary is provided in figure 5.1.

1. *Improvements in device design.*

   The following features are essential to enable the patients to effortlessly embrace wearable devices to improve their quality of life (QoL).

   - *Minimal design*: Devices for neurocognitive conditions are usually targeted towards the older generation. Therefore, the design should be kept minimal. Reducing the number of sensors, keeping the device lightweight, and incorporating unobtrusive elements are a few ways to keep the design minimal and encourage regular wear. More research is needed to determine the optimal number of sensors and optimal sensor placement required to diagnose or monitor the condition. The sensor placement should allow the users to perform daily activities without restrictions.

- *Low cost*: The device cost should not be prohibitive enough to cause a bias in who can potentially use the device.
- *Portability*: The device should be easily usable outdoors during the user's normal daily activities without any discomfort. Smartphone-based and wrist-based wearables are examples of portable devices.
- *User-friendly interface*: The monitor or device should have a user-friendly interface that could enable the users to share and keep in contact with the healthcare provider via a telemedicine-like set-up. This regular exchange of information is critical for remote monitoring and permits symptom management and treatment adjustments.
- *Passive data gathering*: The wearable devices should function well with the least number of alerts and user interactions. The device should also look and feel socially acceptable. A device that garners unwanted attention could disclose the patient's condition, lead to discrimination and social stigma, and act as a constant self-reminder of the user's state.
- *Confidentiality*: The overall design should also follow the laws of patient privacy and security. Data confidentiality is a serious concern and should be addressed in all stages of data flow—data collection, storage, analysis, and result reporting. Transparency of the details around patient privacy and security practices could also alleviate the patient's concerns.
- *Support*: Adequate technical support should be established to address any malfunctions. There must also be an easy maintenance and repair plan in place. Clinical support is also crucial, and both the users and physicians should receive adequate training to use the device seamlessly.
- *Increase value*: If possible, information from the existing sensors in the device should also be effectively utilized to monitor other parameters that could add value to managing the specific condition and monitoring other co-existing conditions. For example, a smartphone app that analyzes data from a wearable that monitors study gait disturbances could also potentially utilize the heart rate information captured by the smartphone's in-built sensors and study correlations between the parameters.
- *Smart homes*: Another area in the pervasive healthcare realm for neurocognitive disorder management is the design and development of smart homes (SHs). Most patients with neurocognitive disorders prefer to stay in their homes or communities and age in place. Aging in place has been shown to reduce the disease's progression (Hyde *et al* 2015). An SH environment is equipped with a collection of interconnected hardware components such as imaging sensors, actuators, gateways, in-home interfaces such as smartphones, TVs, tablets, and software components such as communication interfaces and reasoning

engines. While monitoring the subject, the SH can provide varying interventions such as suggesting actions, prompting or enforcing actions, and performing activities on behalf of the subjects if a risky situation is unraveling (Amiribesheli and Bouchachia 2018). Utilizing such SHs will have a profound effect on elderly care. However, these SHs need further development in personalization for a particular patient among many in nursing care, standardization of evaluation frameworks, and design considering ethical, clinical, and economic aspects of the deployment region.

2. *Large-scale clinical trials.*

   Research shows that several pervasive devices are being designed and tested, but not many gain regulatory approvals and enter the clinical settings and consumer market. The key reason is the lack of large-scale standardized clinical trials. Longitudinal, large-scale observational studies are needed to validate these devices by studying inter and intrasubject variabilities and population variance. The clinical trials should also enroll age-matched controls, and disease-stage matched participants. Studies for a specific condition should also have standardized guidelines so that all clinical studies are easily comparable. The US FDA recently issued guidelines for devices with digital health content to encourage the development and appropriate validation of potentially impactful digital health devices (FDA 2021). Moreover, analysis should be conducted on more real-world data (RWD) collected during the routine use of the devices instead of from standard clinical trials. Analysis of large-scale RWD will result in insightful real-world evidence (RWE) related to the use of the devices in home settings (Rudrapatna and Butte 2020).

3. *Use of AI and predictive analytics.*

   Predictive analytics techniques are being explored extensively to improve the computational models for disease management. An important aspect of predictive modeling is feature engineering, which is the analysis conducted to determine the best features or parameters of a particular disease that provide valuable insight into the illness, enabling better diagnosis, management, and treatment. In neurocognitive disorders, a significant concern is to successfully differentiate signs of aging from the actual symptoms of disorders. In the early stages of the disease, this concern is genuine. Predictive modeling with large volumes of diverse data could provide more insights into the subtle physical and psychological changes that could differentiate aging symptoms from symptoms of early-stage disorders. This type of approach could also enable the discovery of more disease-related digital biomarkers useful for objective evaluation of the patient's health and population health concerning the disease. These biomarkers could also act as baseline reference standards in large-scale clinical trials that assess the effectiveness of the timing and stage of treatments.

   The predictive models could also be designed to indicate the onset of a symptom and track and report the rate of progression of neurological health deterioration to the physicians so they can perform further assessments or alter treatments. Another important aspect of predictive modeling is determining a comprehensible set of features for clinicians and users to understand why and how they help in the decision-making process. For this reason, it is good practice to first work with domain-specific features and classification algorithms that are more transparent to the end-user. Subsequently, deep learning methodologies could be employed to study large amounts of data and develop personalized models for each patient.

5.5.2 Future directions in the application space

1. *Alzheimer's disease.*

   Kourtis *et al* (2019) review a variety of wearables used for AD. Some of them include microphones, cameras to register eye movements and facial expressions, photoplethysmography to record heart rate variability, IMUs to track activity, gait, and tremors, GPS sensors, thermometers, EMG sensors to assess muscle activity, etc. In addition to the improvements suggested earlier, these pervasive devices could also incorporate better algorithms for AD risk prediction and early diagnosis. For example, each of the above sensors could provide information and metrics independently on a specific condition (e.g. gait or cognition). Predictive algorithms could effectively combine these independent metrics into a single overall neurological health metric that could be a basis for a multivariate risk prediction, detection, and symptom monitoring system (Kourtis *et al* 2019).

   Another approach to building such risk prediction models is to utilize nationwide population cohort large-scale datasets. For example, Park *et al* (2020) developed and tested a data-driven machine learning model based on large-scale administrative health data for AD risk prediction. They trained and tested several machine learning models to predict the 1, 2, 3, and 4 year risk for developing AD in the population studied. They reported reasonable prediction AUCs.
2. *Parkinson's disease.*

   Early detection: Overall, there are not many studies focusing on early detection technology validation for PD. The few studies published in the literature are limited by the nature of the dataset used (second-stage patients enrolled for early detection studies that are not appropriate) and the study's scale. Large-scale screening studies should be designed and conducted in patients with mild impairments to adequately validate any technology designed for early detection.

   Symptom monitoring and rehabilitation: The chapter presented several cueing devices to detect FoG episodes to help patients manage them. However, a much better approach would be to predict the onset of FoG episodes earlier to prevent them from happening. Rovini *et al* (2017) provide a detailed overview of specific experimental, technological, and clinical opportunities yet to be explored in the management of PD with pervasive devices.

3. *Epilepsy*.

   Wearable electronic devices for seizure detection could be valuable for individuals with epilepsy as they directly affect obtaining timely interventions and treatments. Just as for AD and PD, these pervasive devices for epilepsy could also provide more freedom and safety to the users. As presented earlier, there are more opportunities for improvements in the device design and algorithms. In terms of epilepsy, the following areas are ripe with the need for improvement:

   - *Real-time prediction and detection*: The devices embedded with real-time automated detection algorithms that can predict the onset of seizures are more valuable. In terms of detection itself, the devices should have a detection sensitivity of above 90% and a false alarm rate of up to one per week (Bruno *et al* 2020).
   - *Seizure localization*: There is a need to develop robust machine learning and deep learning predictive algorithms to predict the onset and detect seizures in long-duration EEG recordings from all EEG channels. Analysis of large-scale multi-electrode EEG datasets could also provide an insight into seizure localization and type. This type of information will be valuable to neurosurgeons (Siddiqui *et al* 2020).

References

Alzheimer's Association 2020 Alzheimer's disease facts and figures *Alzheimer's Dement.* **16** 391–460

Ahn D, Chung H, Lee H W, Kang K, Ko P W, Kim N S and Park T 2017 Smart gait-aid glasses for Parkinson's disease patients *IEEE Trans. Biomed. Eng.* **64** 2394–402

Albers M W *et al* 2015 At the interface of sensory and motor dysfunctions and Alzheimer's disease *Alzheimer's Dement.* **11** 70–98

Amiribesheli M and Bouchachia H 2018 A tailored smart home for dementia care *J. Ambient Intell. Humaniz. Comput.* **9** 1755–82

Arends J *et al* 2018 Multimodal nocturnal seizure detection in a residential care setting: a long-term prospective trial *Neurology* **91** e2010–9

Azulay J P, Mesure S, Amblard B, Blin O, Sangla I and Pouget J 1999 Visual control of locomotion in Parkinson's disease *Brain* **122** 111–20

Bächlin M, Plotnik M, Roggen D, Maidan I, Hausdorff J M, Giladi N and Tröster G 2010 Wearable assistant for Parkinson's disease patients with the freezing of gait symptom *IEEE Trans. Inf. Technol. Biomed.* **14** 436–46

Barthel C *et al* 2018 The laser shoes: a new ambulatory device to alleviate freezing of gait in Parkinson disease *Neurology* **90** e164–71

Beniczky S, Conradsen I, Henning O, Fabricius M and Wolf P 2018 Automated real-time detection of tonic-clonic seizures using a wearable EMG device *Neurology* **90** e428–34

Beniczky S, Polster T, Kjaer T W and Hjalgrim H 2013 Detection of generalized tonic-clonic seizures by a wireless wrist accelerometer: a prospective, multicenter study *Epilepsia* **54** e58–61

Biondi J, Fernandez G, Castro S and Agamennoni O 2017 Eye-movement behavior identification for AD diagnosis arXiv:1702.00837

Bliwise D L 2004 Sleep disorders in Alzheimer's disease and other dementias *Clin. Cornerstone.* **6** S16–28

Boe A J, McGee Koch L L, O'Brien M K, Shawen N, Rogers J A, Lieber R L, Reid K J, Zee P C and Jayaraman A 2019 Automating sleep stage classification using wireless, wearable sensors *NPJ Digit. Med.* **2** 131

Bruno E, Viana P F, Sperling M R and Richardson M P 2020 Seizure detection at home: do devices on the market match the needs of people living with epilepsy and their caregivers? *Epilepsia* **61** S11–24

Bryant M S, Rintala D H, Lai E C and Protas E J 2010 A pilot study: influence of visual cue color on freezing of gait in persons with Parkinson's disease *Disabil. Rehabil. Assist. Technol.* **5** 456–61

Buated W, Sriyudthsak M, Sribunruangrit N and Bhidayasiri R 2012 A low-cost intervention for improving gait in Parkinson's disease patients: a cane providing visual cues *Eur. Geriatr. Med.* **3** 126–30

Bunting-Perry L, Spindler M, Robinson K M, Noorigian J, Cianci H J and Duda J E 2013 Laser light visual cueing for freezing of gait in Parkinson disease: a pilot study with male participants *J. Rehabil. Res. Dev.* **50** 223–30

Castaneda D, Esparza A, Ghamari M, Soltanpur C and Nazeran H 2018 A review on wearable photoplethysmography sensors and their potential future applications in health care *Int. J. Biosens. Bioelectron.* **4** 195–202

Channa A, Popescu N and Ciobanu V 2020 Wearable solutions for patients with Parkinson's disease and neurocognitive disorder: a systematic review *Sensors* **20** 2713

Chen P-H, Wang R-L, Liou D-J and Shaw J-S 2013 Gait disorders in Parkinson's disease: assessment and management *Int. J. Gerontol.* **7** 189–93

Cohen J A, Verghese J and Zwerling J L 2016 Cognition and gait in older people *Maturitas* **93** 73–7

Cova I, Markova A, Campini I, Grande G, Mariani C and Pomati S 2017 Worldwide trends in the prevalence of dementia *J. Neurol. Sci.* **379** 259–60

Daneault J F, Carignan B, Codère C É, Sadikot A F and Duval C 2013 Using a smart phone as a standalone platform for detection and monitoring of pathological tremors *Front. Hum. Neurosci.* **6** 357

de Zambotti M, Rosas L, Colrain I M and Baker F C 2019 The sleep of the ring: comparison of the ŌURA sleep tracker against polysomnography *Behav. Sleep Med.* **17** 124–36

Dietz M A, Goetz C G and Stebbins G T 1990 Evaluation of a modified inverted walking stick as a treatment for Parkinsonian freezing episodes *Mov. Disord.* **5** 243–7

Donovan S, Lim C, Diaz N, Browner N, Rose P, Sudarsky L R, Tarsy D, Fahn S and Simon D K 2011 Laserlight cues for gait freezing in Parkinson's disease: an open-label study *Parkinsonism Relat. Disord.* **17** 240–5

Doraiswamy P M, Narayan V A and Manji H K 2018 Mobile and pervasive computing technologies and the future of Alzheimer's clinical trials *NPJ Digit. Med.* **1** 1

Elias W J and Shah B B 2014 Tremor *JÅMÅ* **311** 948–54

Enzensberger W, Oberländer U and Stecker K 1997 Metronomtherapie bei Parkinson-Patienten [Metronome therapy in patients with Parkinson disease] *Nervenarzt* **68** 972–7

Espay A J, Baram Y, Dwivedi A K, Shukla R, Gartner M, Gaines L, Duker A P and Revilla F J 2010 At-home training with closed-loop augmented-reality cueing device for improving gait in patients with Parkinson disease *J. Rehabil. Res. Dev.* **47** 573–81

FDA 2021 https://fda.gov/medical-devices/digital-health-center-excellence/guidances-digital-health-content

Fernández G, Mandolesi P, Rotstein N P, Colombo O, Agamennoni O and Politi L E 2013 Eye movement alterations during reading in patients with early Alzheimer disease *Invest. Ophthalmol. Vis. Sci.* **54** 8345–52

Fleming S, Thompson M, Stevens R, Heneghan C, Plüddemann A, Maconochie I, Tarassenko L and Mant D 2011 Normal ranges of heart rate and respiratory rate in children from birth to 18 years of age: a systematic review of observational studies *Lancet* **377** 1011–8

Giladi N, Horak F B and Hausdorff J M 2013 Classification of gait disturbances: distinguishing between continuous and episodic changes *Mov. Disord.* **28** 1469–73

Giuffrida J P, Riley D E, Maddux B N and Heldman D A 2009 Clinically deployable Kinesia technology for automated tremor assessment *Mov. Disord.* **24** 723–30

Hyde J, Perez R, Doyle P J, Forester B P and Whitfield T H 2015 The impact of enhanced programming on aging in place for people with dementia in assisted living *Am. J. Alzheimer's Dis. Other Demen.* **30** 733–7

Jian M A 2020 Associations between finger tapping, gait and fall risk with application to fall risk assessment arXiv: 2006.16648

Joshi R *et al* 2019 PKG movement recording system use shows promise in routine clinical care of patients with Parkinson's disease *Front. Neurol.* **10** 1027

Kourtis L C, Regele O B, Wright J M and Jones G B 2019 Digital biomarkers for Alzheimer's disease: the mobile/wearable devices opportunity *NPJ Digit. Med.* **2** 9

Lan K-C and Shih W-Y 2014 Early diagnosis of Parkinson's disease using a smartphone *Procedia Comput. Sci.* **34** 305–12

Lim A S, Kowgier M, Yu L, Buchman A S and Bennett D A 2013 Sleep fragmentation and the risk of incident Alzheimer's disease and cognitive decline in older persons *Sleep* **36** 1027–32

Lonini L, Dai A, Shawen N, Simuni T, Poon C, Shimanovich L, Daeschler M, Ghaffari R, Rogers J A and Jayaraman A 2018 Wearable sensors for Parkinson's disease: which data are worth collecting for training symptom detection models *NPJ Digit. Med.* **1** 64

Manson A, Stirpe P and Schrag A 2012 Levodopa-induced-dyskinesias clinical features, incidence, risk factors, management and impact on quality of life *J. Parkinson's Dis.* **2** 189–98

Maresova P, Tomsone S, Lameski P, Madureira J, Mendes A, Zdravevski E, Chorbev I, Trajkovik V, Ellen M and Rodile K 2018 Technological solutions for older people with Alzheimer's disease: review *Curr. Alzheimer Res.* **15** 975–83

Margiotta N, Avitabile G and Coviello G 2017 A wearable wireless system for gait analysis for early diagnosis of Alzheimer and Parkinson disease *Int. Conf. on Electronic Devices, Systems, and Applications*

Mazilu S, Blanke U, Dorfman M, Gazit E, Mirelman A, Hausdorff J M and Tröster G 2015 A wearable assistant for gait training for Parkinson's disease with freezing of gait in out-of-the-lab environments *ACM Trans. Interact. Intell. Syst.* **15** 1

McCandless P J, Evans B J, Janssen J, Selfe J, Churchill A and Richards J 2016 Effect of three cueing devices for people with Parkinson's disease with gait initiation difficulties *Gait Posture* **44** 7–11

Meritam P, Ryvlin P and Beniczky S 2018 User-based evaluation of applicability and usability of a wearable accelerometer device for detecting bilateral tonic-clonic seizures: a field study *Epilepsia* **59** 48–52

Ngueleu A M *et al* 2019 Validity of instrumented insoles for step counting, posture and activity recognition: a systematic review *Sensors* **19** 2438

Nieuwboer A *et al* 2007 Cueing training in the home improves gait-related mobility in Parkinson's disease: the RESCUE trial *J. Neurol. Neurosurg. Psychiatry* **78** 134–40

Nonnekes J, Snijders A H, Nutt J G, Deuschl G, Giladi N and Bloem B R 2015 Freezing of gait: a practical approach to management *Lancet Neurol.* **14** 768–78

Park J H, Cho H E, Kim J H, Wall M M, Stern Y, Lim H, Yoo S, Kim H S and Cha J 2020 Machine learning prediction of incidence of Alzheimer's disease using large-scale administrative health data *NPJ Digit Med.* **3** 46

Perumal S V and Sankar R 2016 Gait and tremor assessment for patients with Parkinson's disease using wearable sensors *ICT Express* **2** 168–74

Poh M Z, Loddenkemper T, Swenson N C, Goyal S, Madsen J R and Picard R W 2010 Continuous monitoring of electrodermal activity during epileptic seizures using a wearable sensor *Annu. Int. Conf. IEEE Eng. Med. Biol. Soc.* (Piscataway, NJ: IEEE) pp 4415–8

Rabinowitz I and Lavner Y 2014 Association between finger tapping, attention, memory, and cognitive diagnosis in elderly patients *Percept. Mot. Skills* **119** 259–78

Ramdhani R A, Khojandi A, Shylo O and Kopell B H 2018 Optimizing clinical assessments in Parkinson's disease through the use of wearable sensors and data driven modeling *Front. Comput. Neurosci.* **12** 72

Regalia G, Onorati F, Lai M, Caborni C and Picard R W 2019 Multimodal wrist-worn devices for seizure detection and advancing research: focus on the Empatica wristbands *Epilepsy Res.* **153** 79–82

Renevey P, Delgado-Gonzalo R, Lemkaddem A, Proença M, Lemay M, Solà J, Tarniceriu A and Bertschi M 2017 Optical wrist-worn device for sleep monitoring *European Medical and Biological Engineering Conf., Nordic-Baltic Conf. on Biomedical Engineering and Medical Physics* pp 615–8

Rentz D M *et al* 2016 The feasibility of at-home iPad cognitive testing for use in clinical trials *J. Prev. Alzheimer's Dis.* **3** 8–12

Rigas G, Tzallas A T, Tsipouras M G, Bougia P, Tripoliti E E, Baga D, Fotiadis D I, Tsouli S G and Konitsiotis S 2012 Assessment of tremor activity in the Parkinson's disease using a set of wearable sensors *IEEE Trans. Inf. Technol. Biomed.* **16** 478–87

Rosenthal L, Sweeney D, Cunnington A L, Quinlan L R and ÓLaighin G 2018 Sensory electrical stimulation cueing may reduce freezing of gait episodes in Parkinson's disease *J. Healthc. Eng.* **2018** 4684925

Rovini E, Maremmani C and Cavallo F 2017 How wearable sensors can support Parkinson's disease diagnosis and treatment: a systematic review *Front. Neurosci.* **11** 555

Rudrapatna V A and Butte A J 2020 Opportunities and challenges in using real-world data for health care *J. Clin. Invest.* **130** 565–74

Rukasha T, Woolley I, Kyriacou S and Collins T 2020 Evaluation of wearable electronics for epilepsy: a systematic review *Electronics* **9** 968

Santiago A *et al* 2019 Qualitative evaluation of the personal KinetiGraphTM movement recording system in a Parkinson's clinic *J. Parkinson's Dis.* **9** 207–19

Sheng J, Liu S, Qin H, Li B and Zhang X 2018 Drug-resistant epilepsy and surgery *Curr. Neuropharmacol.* **16** 17–28

Siddiqui M K, Morales-Menendez R, Huang X and Hussain N 2020 A review of epileptic seizure detection using machine learning classifiers *Brain Inform.* **7** 5

Stringer G *et al* 2018 Can you detect early dementia from an email? A proof of principle study of daily computer use to detect cognitive and functional decline *Int. J. Geriatr. Psych.* **33** 867–74

Sweeney D, Quinlan L R, Browne P, Richardson M, Meskell P and ÓLaighin G 2019 A technological review of wearable cueing devices addressing freezing of gait in Parkinson's disease *Sensors* **19** 1277

Tang L, Gao C, Wang D, Liu A, Chen S and Gu D 2017 Rhythmic laser cue is beneficial for improving gait performance and reducing freezing of turning in Parkinson's disease patients with freezing of gait *Int. J. Clin. Exp. Med.* **10** 16802–8

Van Andel J, Gutter T, Ungureanu C, Arends J and Leijten F 2015 A comparison of heart rate patterns in clinically urgent and non-urgent nocturnal motor seizures *Towards a Multimodal System for Nocturnal Seizure Detection* ed J Van Andel (Utrecht: University of Utrecht)

WHO 2019 *Epilepsy: A Public Health Imperative* (Geneva: World Health Organization)

Zhao Y *et al* 2016 Feasibility of external rhythmic cueing with the Google Glass for improving gait in people with Parkinson's disease *J. Neurol.* **263** 1156–65

IOP Publishing

Predictive Analytics in Healthcare, Volume 1
Transforming the future of medicine
Vinithasree Subbhuraam

Chapter 6

Extraction of fetal head section from ultrasound images using a soft-computing based image mining system—a study with Kapur's thresholding and segmentation

V Rajinikanth and Seifedine Kadry

Biomedical image evaluation is a promising field due to its importance and clinical significance. This research work aims to develop an image mining system to extract a fetal head section (FHS) from an ultrasound image (UI) using a semi-automated scheme with a soft-computing technique. The extraction and measurement of the fetal head circumference (FHC) are essential for monitoring the progress of the fetus. The proposed approach implements an integration of an image thresholding and segmentation scheme to extract the FHC from the benchmark HC18 database. The initial enhancement of the test image is performed using tri-level thresholding with Kapur's function and the particle swarm optimization (PSO) algorithm. The enhanced FHS is then mined using a chosen semi-automated segmentation technique. The semi-automated segmentation techniques, such as level set, active contour, and Chan–Vese are executed to extract the FHS and the performance of the considered methods is validated based on the obtained values of the Jaccard index (JI), Dice's coefficient (DC), and segmentation accuracy. One hundred test images of the HC18 dataset are considered for the experimental demonstration and the result of the proposed work confirms that the implemented technique helps to obtain better values of JI (90.27%), DC (93%), and segmentation accuracy (95.12%) with the active-contour segmentation technique. The performance of the PSO is also validated with the Bat algorithm (BA) and firefly algorithm (FA) available in the literature and the PSO presents a better convergence compared to the alternatives.

doi:10.1088/978-0-7503-2312-3ch6

6.1 Introduction

In the current era, due to improvements in science and engineering, considerable advancements are occurring in every domain, including the medical field. Recent advancements in the medical domain such as diagnosing diseases in their early phase, appropriate treatment implementation, remote patient monitoring, telemedicine, etc, have helped the human community in achieving quality medical care [1–5].

The diseases and conditions of the human physiological system can be monitored effectively using various procedures, including biomedical signals/images collected using a variety of modalities [6–11]. Biomedical images will provide complete information regarding the section to be examined compared to bio-signals, and hence several imaging modalities have been invented and used in the medical domain to evaluate the condition of different body parts. Most biomedical imaging modalities were invented to support the non-invasive assessment of internal and external body organs to identify their functioning in various situations. Bio-imaging methods, such as computer tomography (CT), magnetic resonance imaging (MRI), and x-rays require radiographic waves to capture the functioning of internal organs and these imaging methods are recommended by the doctor only when it is essential.

Along with the radiographic bio-imaging methods, other imaging methods such as thermal imaging and ultrasound imaging are also adopted widely to diagnose the condition of internal organs due to their simple and non-invasive nature. These methods are employed widely in the medical domain to record the condition of various internal organs, including fetal development. The employment of a selected biomedical imaging practice begins at the premature phase of the fetus, and the stages with respect to time and the health of fetal growth are assessed regularly using the imaging modality called ultrasound/sonography. In this technique the activity and other information on the fetus is recorded and evaluated using a high-frequency sound wave.

Prenatal viewing is one of the familiar procedures for prenatal diagnosis and it is achieved with obstetric ultrasonography, or prenatal ultrasound [12–14]. This process helps to record fetal growth during pregnancy using ultrasound-assisted imaging. After confirming the pregnancy, continuous/scheduled testing is necessary to identify abnormalities during the fetus's growth. Implementing an ultrasound assessment at this premature stage of pregnancy can assist in accurately confirming the timing of the pregnancy, and can also look for multiple fetuses and major congenital abnormalities at a early stage.

The common procedure followed in prenatal diagnosis is to estimate fetal growth by simply measuring the head circumference. Normally, examination of the fetal growth between predefined periods, such as 28–32 weeks, is commonly recommended and this scan can be performed once or many times based on the guidance of the doctor. The common recording procedure followed in hospitals is depicted in figure 6.1, in which a clinically accepted ultrasound scanner (transmitter and receiver) is used to record the orientation and the activities of the fetus. The captured image can be monitored using a display unit or a recorded image on a

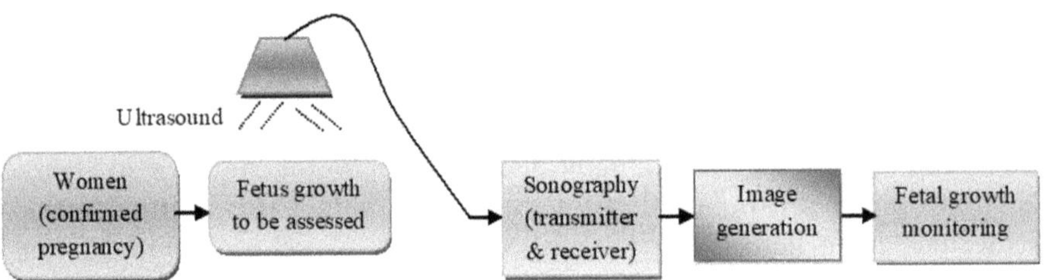

Figure 6.1. Fetus growth monitoring using ultrasound imaging.

transparent film. The assessment will help to identify birth defects and other complications at the initial stage and, if needed, the physician may recommend specific steps to maintain fetal growth.

Due to its clinical significance, a considerable number of sonography based fetus growth detection procedures have been discussed and proposed widely by researchers. Several studies have established that the detection and assessment of fetal head circumference (FHC) are essential to predict the growth rate [15–18]. In these studies a considerable number of procedures have been proposed and implemented to extract and evaluate the FHC using several image processing techniques. In this chapter, a heuristic algorithm based image processing procedure is proposed and implemented to extract and evaluate the FHC section from the benchmark ultrasound FHC dataset available in [19, 20].

The proposed image processing technique helps to implement the following methods: (i) an image pre-processing technique to enhance the FHC from the considered image slice, (ii) an image post-processing technique to extract the FHC from the pre-processed image, and (iii) comparison and validation of the extracted FHC with the ground truth (GT) provided by an expert. In this work, the initial pre-processing implements a tri-level thresholding technique to separate the test image into three sections, namely the background, FHC, and other sections based on the pixel group. This image thresholding is implemented using Kapur's/Otsu's technique and the essential threshold value is identified using the particle swarm optimization (PSO) algorithm. The image post-processing is implemented using a chosen segmentation methodology and in this chapter the semi-automated segmentation methods level set (LS), active contour (AC), and Chan–Vese (CV) are implemented and the performances of these segmentation methods are compared. Finally, a comparative assessment between the extracted FHS and the GT is implemented and the essential image information is computed. The dominance of the implemented thresholding and the segmentation procedure is then validated based on the computed values of the Jaccard index (JI), Dice's coefficient (DC), and segmentation accuracy.

The remaining parts of the chapter are organized as follows: section 6.2 discusses the context and reviews the relevant past literature in this area, section 6.3 presents the methodology, the results are presented in section 6.4, the discussion is given in section 6.5, and the conclusion is provided in section 6.6.

6.2 Context

Even though a considerable number of different imaging modalities are available, ultrasound imaging is adopted widely in the medical domain due to its simplicity and non-invasive nature. Previously, it was commonly used to diagnose the various conditions of the fetus. Further, recently it has been adopted to detect abnormalities in various organs, including abnormal breast tissue [1, 21]. Once the basic ultrasound image of the fetus is acquired using appropriate medical guidelines, the chosen image processing technique is then employed to extract and evaluate the FHC from a grayscale image of chosen dimensions. The assessment of the fetus using sonography is a recommended methodology and table 6.1 summarizes the existing methods in the literature.

Table 6.1. Lung abnormality assessment using computerized approaches.

Reference	Methodology implemented
Hadlock *et al* [22]	This study assessed fetal head to abdomen circumference measurements using ultrasound techniques using 568 images collected from patients with normal pregnancies from 17 to 41 weeks. The computed mean ratio was 1.18 at 17 weeks and this ratio gradually decreased with an increase in fetal growth.
Valsky *et al* [23]	This study conducted a detailed examination of the head circumference (HC), birth weight, second stage duration, maternal age, baby sex, episiotomy, and instrumental delivery.
Elvander *et al* [24]	This study implemented a detailed examination of various head circumferences of newborn babies and presented a detailed analysis of the head circumference with the labor outcome.
Lipschuetz *et al* [25]	Ultrasound image assisted assessment of the fetal head circumference and estimated fetal weight were presented using various images collected from normal pregnancies of 37–42 weeks.
Van *et al* [20]	This study proposed a computer-aided detection (CAD) system to assess the FHC in 2D ultrasound images collected from the HC18 Grand Challenge database. This work implemented an image processing methodology to extract and evaluate the FHC from 1334 images (999 for training and 335 for validation) and attained better segmentation results.
Li *et al* [26]	This study conducted an image-based assessment of fetal growth during prenatal ultrasound examinations using random forest and fast ellipse fitting methodologies and the implemented technique evaluated the FHC of 145 images collected using a medical protocol.
Sobhaninia *et al* [27]	This work implemented a multi-task deep learning system to extract the FHC from fetal ultrasound images and extracted and evaluated the condition of the fetus based on the computed biometric parameters.
Rajinikanth *et al* [28]	This work implemented a hybrid image processing procedure to extract and evaluate the FHC from ultrasound images. They integrated image thresholding with image segmentation, compared the extracted FHS against the GT, and then computed fundamental image performance measures.

From table 6.1 it can be observed that a considerable number of methods have already been proposed and implemented in the literature to extract and evaluate the FHC section. The recent work of Rajinikanth *et al* [28] reported that the hybrid imaging methodology implemented using the image pre-processing and post-processing helped to achieve an overall performance >88%. This value can be further improved by considering other suitable image processing techniques.

In the proposed technique, a detailed assessment of image thresholding methods, namely Kapur and Otsu, is presented initially using a chosen heuristic algorithm and then the FHC section from the thresholded image is extracted using well-known segmentation techniques. Finally, the extracted FHC from every chosen technique is compared against the GT and the fundamental image quality measures are computed. Based on the attained values, the performance of the implemented procedure is then confirmed.

6.3 Methodology

Prenatal viewing is an essential procedure to continuously monitor the growth rate and health of the fetus and FHC based assessment is one of the widely adopted techniques. In the literature, a considerable number of image evaluation techniques have been proposed and implemented to assess the FHC from 2D ultrasound images. In a recent work, Rajinikanth *et al* [28] confirmed that hybrid image examination procedures will offer a better evaluation result for FHC evaluation. The work proposed in this chapter aims to provide a detailed evaluation of the (i) chosen heuristic algorithms, (ii) image thresholding techniques, and (iii) segmentation procedures.

Figure 6.2 depicts the various stages of the proposed FHC evaluation technique. The initial images for the assessment are collected from the HC18 Grand Challenge database. Initially, the existing test image and the GT are resized into $512 \times 512 \times 1$ pixels and the resized images are then considered for the assessment. The resized image is then pre-processed using a tri-level thresholding technique, which helps to group the test image pixels into three sections, namely background, head section, and other body parts using an Kapur/Otsu based technique. To identify the optimal threshold for the test image, the well-known heuristic algorithms PSO, BA, and FA are considered.

The enhanced head section from the thresholded image is then extracted using a chosen segmentation procedure. The segmentation procedure helps to provide a binary image and this image is then compared with the GT and the fundamental image quality measures are computed. Based on the computed performance values, the performance of the proposed technique is validated and confirmed.

6.3.1 Ultrasound image database

Prenatal viewing is an approved medical procedure that is performed widely to assess the normal growth of the fetus using a suitable imaging technique. Due to its efficiency and radiation-free nature, sonography is commonly used to examine the growth rate of the fetus. Ultrasounds are generally performed by experienced

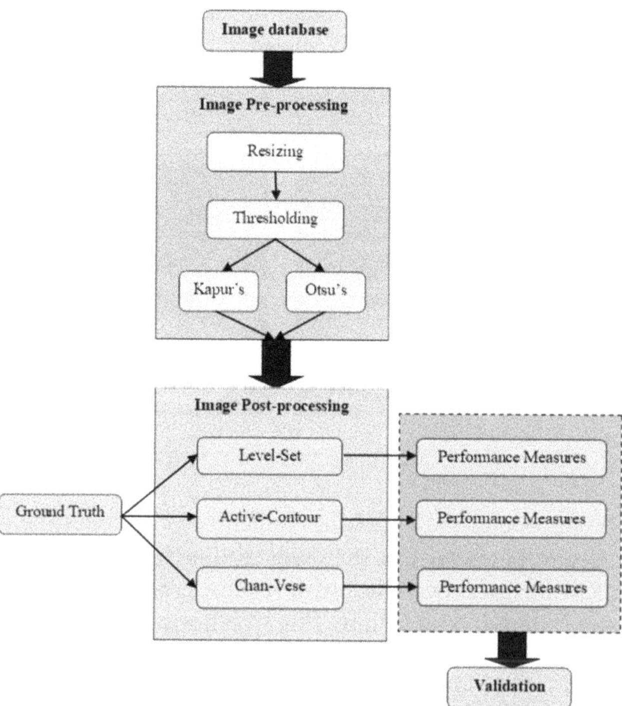

Figure 6.2. Various phases in the proposed FHC extraction and evaluation process. Reproduced with permission from HC18 Grand Challenge, CC 4.0.

radiologists with guidance from physicians. In addition to image acquisition, image interpretation plays a vital role in accurate diagnosis. To achieve an effective diagnosis, a considerable number of image assisted assessment procedures have been proposed and implemented in the literature.

The availability of fetus ultrasound images for research is very limited due to various ethical and copyright issues, and obtaining a clinical-grade image for assessment is a challenging task. The HC18 Grand Challenge database consists of 1334 clinical-grade 2D ultrasound images, which can be used for research without restrictions. This dataset consists of 999 training images along with the GT and 335 testing images. The original test image and the GT are available with dimensions of $800 \times 540 \times 1$ and the sample test images and the associated GT are depicted in figure 6.3. This figure presents the test images and the actual GT available in the database and the synthetic GT generated using the ITK-Snap tool [29, 30].

The aim of the proposed research in this chapter is to develop a computerized tool to extract the head section from the resized test image and the extracted section is then compared against the binary GT to confirm the performance of the proposed technique.

6.3.2 Pre-processing

Image pre-processing is one of the vital image correction/enhancement techniques commonly adopted in general and medical image enhancement tasks. Multi-level

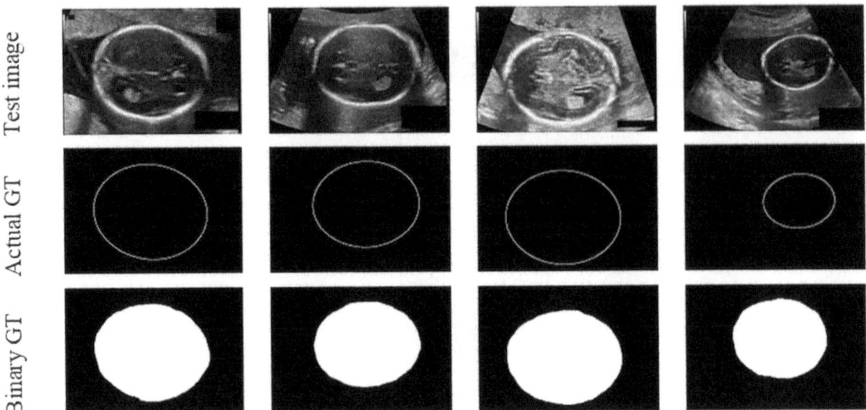

Figure 6.3. The sample test images and the associated GT.

image thresholding is one of the widely implemented pre-processing techniques used to enhance the region of interest in the medical image under examination [31–35]. During the medical image processing, tri-level thresholding is employed widely by researchers to separate the image into three sections based on the pixel groups. Recently, heuristic algorithm based image thresholding has been employed widely by researchers to simplify the computational effort. In this work, the pre-processing is initiated with chosen heuristic algorithms.

6.3.3 Heuristic algorithms

Recently, heuristic algorithm (HA) assisted optimization techniques have been adopted widely to solve a variety of real-world problems and in this work the HA has been considered to find the optimal threshold based on the assigned thresholding function. Even though a considerable number of HAs exist in the literature, this work employed some of the most successful heuristic techniques such as PSO [36, 37], BA [38–40], and FA [41, 42]. In this work, along with the traditional FA (FA1), recent advancements such as the Brownian walk based FA (FA2) and Ikeda map assisted FA (FA3) are considered for the assessment. Information regarding the PSO, BA, and FA can be found in earlier research works. The main aim of the proposed work is to implement the PSO algorithm to perform the optimization task and compare the optimization performance of the PSO with the BA and FA algorithms. The basic information on the considered heuristic methods can be found in the literature [43–45].

6.3.4 Multi-thresholding

Pre-processing of the test image is essential to enhance the image information for further assessment. Even though a considerable number of enhancement techniques are available, thresholding with a chosen number of thresholds is employed widely in the image processing domain to enhance the test image (greyscale/RGB) based on the pixel grouping technique. In this work, the essential thresholding process is implemented using Kapur's and Otsu's techniques, and the performances of these two procedures are confirmed with the final result obtained with the proposed technique.

6.3.4.1 Kapur's approach

Kapur's function has been proposed primarily for thresholding grayscale images using the histogram's entropy [46]. This technique finds the optimal threshold by maximizing the entropy. For an image, if $f(k)$ denotes the frequency of the kth intensity-level, then the pixel distribution of the image will be

$$Z = f(0) + f(1) + \ldots + f(L-1). \tag{6.1}$$

If the probability of the kth intensity-level is given by

$$p(k) = f(k)/Z \tag{6.2}$$

then, during the threshold selection, the pixels of the image are separated into Th + 1 groups according to the assigned threshold value. After extrication of the images as per the selected threshold, the entropy of each cluster is calculated independently and combined to obtain the final entropy as follows:

$$\text{Multi-level threshold} = f(t_1, t_2, \ldots t_L) = e_0 + e_1 + \ldots + e_{L-1} \tag{6.3}$$

$$\begin{aligned}
e_0 &= -\sum_{k=0}^{k=t_{1-1}} \frac{p_k}{\sigma_0} \ln \frac{p_k}{\sigma_0}, \quad \sigma_0 = \sum_{k=0}^{k=t_{1-1}} p_k \\
e_1 &= -\sum_{k=t_{1-1}}^{k=t_{1-2}} \frac{p_k}{\sigma_1} \ln \frac{p_k}{\sigma_1}, \quad \sigma_1 = \sum_{k=t_{1-1}}^{k=t_{1-2}} p_k \\
e_{L-1} &= -\sum_{k=t_{L-1}}^{k=t_{L-2}} \frac{p_k}{\sigma_{L-1}} \ln \frac{p_k}{\sigma_{L-1}}, \quad \sigma_{L-1} = \sum_{k=t_{L-1}}^{k=t_{L-2}} p_k,
\end{aligned} \tag{6.4}$$

where e is the entropy, p is the probability distribution, and σ is the probability occurrence,

$$\text{Kapur}_{\max}(T) = \sum_{p=1}^{L-1} H_j^C. \tag{6.5}$$

Further information on Kapur's function can be found in [47–50].

6.3.4.2 Otsu's approach

Otsu's technique was first discussed in 1979 and it works based on a between-class-variance concept and it identifies the finest threshold by maximizing the objective value [51]. The between-class-variance is Otsu's non-parametric threshold selection concept, to be computed by varying the image threshold randomly using a chosen technique. During this process the selection of Thresholding(Th) is essential to convert the raw image into a processed image. Further related information on Otsu's thresholding can be found in previous works [49, 50, 52].

6.3.5 Segmentation

In medical image processing, segmentation techniques are employed widely to extract a particular section from an image using a chosen semi-automated or automated technique. In the proposed work, the segmentation task is to extract the fetal head from the pre-processed 2D ultrasound image. In this work, the well-known semi-automated segmentation procedures LS [53], AC [54], and CV [55, 56] are employed to extract the head section with better accuracy. All are bounding-box based techniques that will adjust their shape until the fundamental section from the image is extracted. In this work the operator initiates a bounding box that is allowed to converge towards the head section when the iteration increases. When this process identifies all the possible pixels of the head section, the operation is terminated and the segmented binary image of the head section is then presented for further assessment. Previous works on the LS, AC, and CV segmentation can be found in [57, 58].

6.3.6 Data evaluation and validation

The performance of the proposed image examination system has to be validated to confirm the clinical significance of the proposed technique. In this work, to confirm the performance of the proposed image examination technique, a comparative analysis between the GT and the extracted section is presented and the fundamental performance values are computed.

The performance measures considered in this work include the following [59, 60]:

$$\text{Jaccard index} = \frac{\text{GTI} \cap \text{SI}}{\text{GTI} \cup \text{SI}} = \frac{\text{TP}}{\text{TP} + \text{FP} + \text{FN}} \tag{6.6}$$

$$\text{DC} = F_1 \text{ score} = \frac{2\,|\,\text{GTI} \cap \text{SI}\,|}{|\,\text{GTI}\,| + |\,\text{SI}\,|} = \frac{2\text{TP}}{2\text{TP} + \text{FP} + \text{FN}} \tag{6.7}$$

$$\text{Accuracy} = \frac{\text{TP} + \text{TN}}{\text{TP} + \text{TN} + \text{FP} + \text{FN}} \tag{6.8}$$

$$\text{Precision} = \frac{\text{TP}}{\text{TP} + \text{FP}} \tag{6.9}$$

$$\text{Sensitivity} = \frac{\text{TP}}{\text{TP} + \text{FN}} \tag{6.10}$$

$$\text{Specificity} = \frac{\text{TN}}{\text{TN} + \text{FP}}, \tag{6.11}$$

where GTI is the ground truth image, SI is the segmented image, and TP, TN, FP, and FN denote true-positive, true-negative, false-positive, and false-negative, respectively, Further, other related measures, such as the FN rate and FP rate, are also computed.

6.4 Results

This section presents the results obtained with the proposed technique on the HC18 database. The basic experimental investigation are implemented using MATLAB software. The proposed work is tested and validated using 100 HC18 dataset images. Initially, the basic images and the related GT are collected from the dataset. After collecting the basic image for the assessment, an image resizing operation is implemented to reduce the image dimensions from 800 × 540 × 1 pixels to 512 × 512 × 1 pixels, and the reduced image is then considered for the examination.

Initially, HA-based tri-level image thresholding is employed on every test image and the fundamental information from the enhanced image is then extracted using a chosen segmentation technique. Before initiating the HA-based thresholding, the following initial algorithm parameters are assigned for the PSO, BA, and FA: the number of artificial agents = 25, search dimension = 3, search iteration = 3000, and termination criterion = maximal iteration or best-attained result.

In this work, Kapur's thresholding is employed initially to enhance the image and then the basic operations are implemented. Later, a similar procedure is repeated with Otsu's technique and the obtained result of Otsu is compared with Kapur's and the performance is validated.

Figure 6.4 depicts the various thresholding results obtained with the proposed approach using Kapur and Otsu's techniques. From figure 6.4(a) it can be noted that the optimization search convergence obtained with the PSO is better compared to the BA and FA. Similar convergence can also be seen with Otsu's technique, as can be seen in figure 6.4(b). Hence, in the proposed work the PSO algorithm is considered to be implemented in the proposed image processing technique to compute the FHC. Figures 6.4(a) and (b) show that the PSO demonstrates a better and quicker convergence compared to the other heuristic techniques and this procedure confirms that the thresholding performance of the PSO is better compared to the BA and FA techniques.

The thresholding result obtained for a chosen test image can be found in figure 6.5. Figure 6.5(a) shows the chosen test image for the assessment, and figures 6.5(b) and (c) depict the actual and the binary GT, respectively. Figures 6.5(d) and (e) depict the thresholding outcomes with the Kapur and Otsu techniques along with PSO. A similar result is obtained for the thresholding using the BA and FA (FA1, FA2, and FA3).

Figure 6.6 depicts the outcome of the segmentation technique employed to extract the fetal head, and all the implemented techniques helped to extract the image section of interest. Figures 6.6(a) to (d) depict the initial bounding box, converged section, and the results obtained with Kapur's technique and Otsu's technique, respectively. From figures 6.6(c) and (d), it can be noted that both the Kapur as well

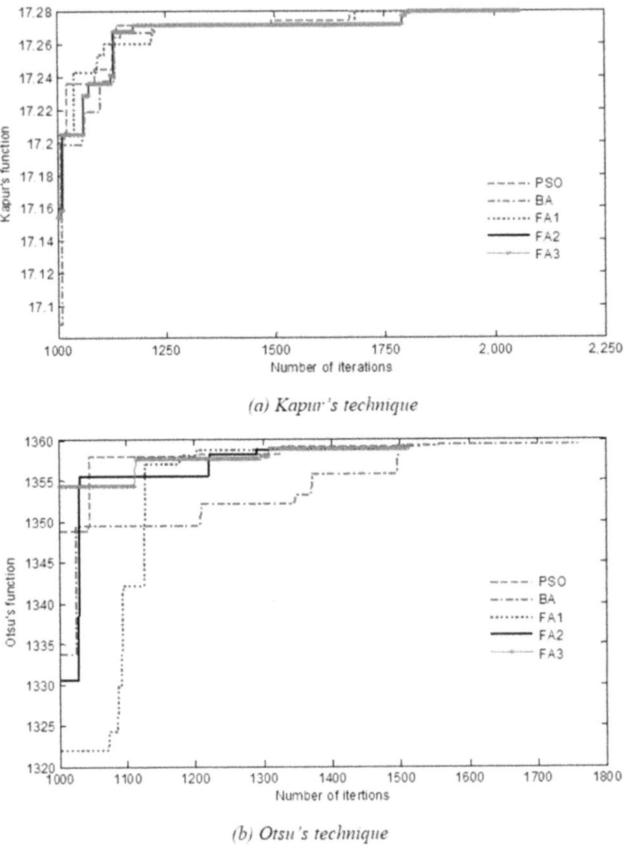

(a) Kapur's technique

(b) Otsu's technique

Figure 6.4. Convergence of the heuristic algorithm based optimization search. Reproduced with permission from HC18 Grand Challenge, CC 4.0.

as the Otsu segmentation helped to obtain a better result on the chosen test image. In this chapter, only Kapur's technique is considered for the demonstration and a similar result is obtained with the Otsu technique.

Tables 6.2 and 6.3 depict the image performance measures computed by comparing the segmented section with the GT. These tables presents the results obtained using the LS, AC, and CV approaches for the PSO and Kapur thresholded images. From tables 6.2 and 6.3, it can be observed that the proposed technique helped to achieve a better result on the chosen test image. The performance measures, namely JI, DC, and the segmentation accuracy, are better (>90%) and hence it can be confirmed that the proposed technique helps to obtain a better result in the considered HC18 database.

Tables 6.2 and 6.3 present the performance measures of the LS, AC, and CV approaches, and the individual performance of these methods as appropriate. To compute the overall performance, a glyph plot is constructed in figure 6.7 and this figure confirms that the segmentation performance of the LS is better compared to the AC and CV techniques. A similar procedure is repeated on the other test images and the obtained results are evaluated and validated. The implementation of the LS

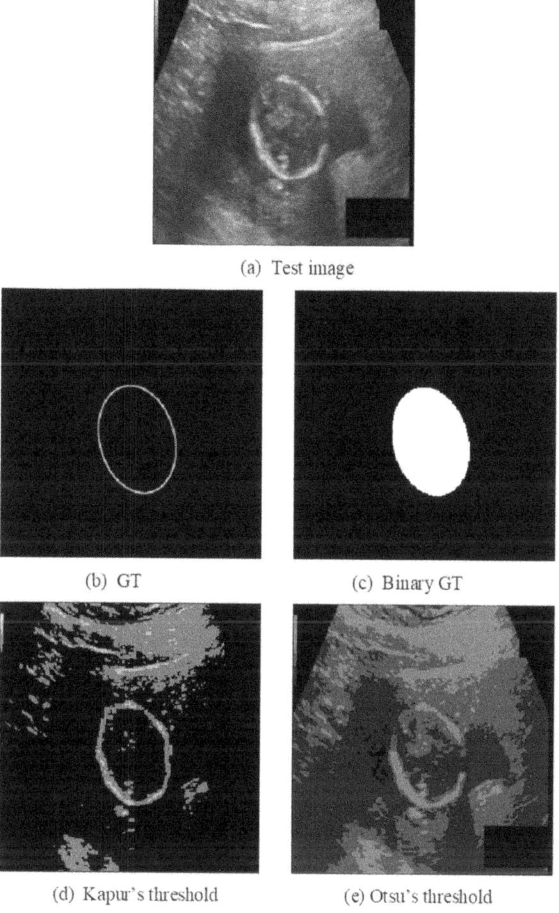

Figure 6.5. Pre-processing result obtained for a chosen ultrasound image. Reproduced with permission from HC18 Grand Challenge, CC 4.0.

is quite similar to the AC and the CV approaches, but the segmentation accuracy obtained by the LS approach is better irrespective of the thresholding outcome obtained with the Kapur or Otsu techniques.

6.5 Discussion

In the proposed study, a computerized image examination is implemented on 2D ultrasound images to extract and evaluate the fetal head section with better accuracy.

This work implemented the PSO algorithm based on Kapur's or Otsu's technique, which enhances the visibility of the test image. Later, a chosen segmentation technique (LS/AC/CV) was implemented to extract the essential region for further assessment. The results obtained for a few resized sample test images are depicted in figure 6.7, and the results confirm the performance of the proposed technique.

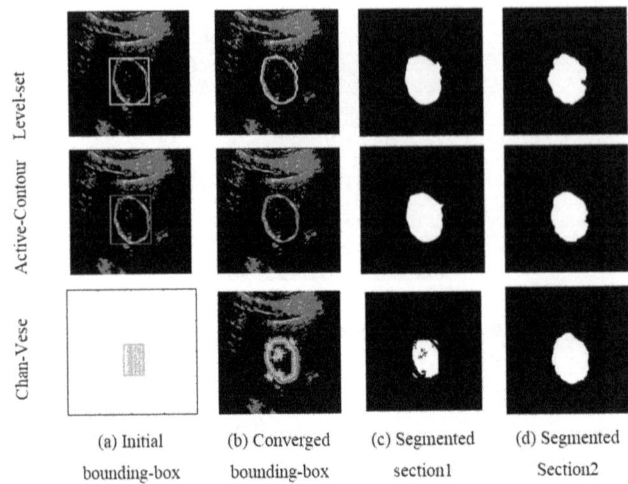

Figure 6.6. Segmentation results obtained with the LS, AC, and CV techniques.

Table 6.2. Performance analysis with Kapur's thresholding and segmentation technique.

Method	TP	FP	TN	FN	FNR	FPR
LS	100 222	499	153 808	7615	0.0706	0.0032
AC	99 311	1410	152 423	9000	0.0831	0.0092
CV	98 807	1914	154 263	7160	0.0676	0.0123

Table 6.3. Image quality measures obtained with the proposed technique.

Method	JI	DC	Accuracy	Precision	Sensitivity	Specificity
LS	0.9251	0.9611	0.9690	0.9950	0.9294	0.9968
AC	0.9051	0.9502	0.9603	0.9860	0.9169	0.9908
CV	0.9159	0.9561	0.9654	0.9810	0.9324	0.9877

Figures 6.8(a) and (b) depict the test images and the associated GT. Figures 6.8(c) to (e) depict the extracted section using the LS, AC, and CV approaches and the results of these figures confirm that the proposed technique helped to provide a better result. This technique was then employed to test all the existing images (n = 100) collected from the database and the corresponding results obtained with the Kapur or Otsu technique is then considered to compute the performance of the image examination system.

Figure 6.9 depicts the performance measures obtained with the PSO algorithm based Kapur or Otsu thresholding and the LS segmentation. A similar methodology is repeated with the AC and the CV segmentation procedures and the results obtained with these approaches confirms that the performance measures obtained

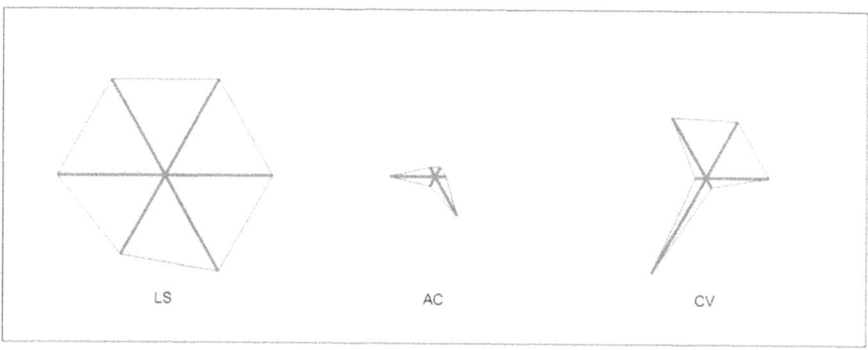

Figure 6.7. Glyph diagram to examine the overall performance of the chosen technique. Reproduced with permission from HC18 Grand Challenge, CC 4.0.

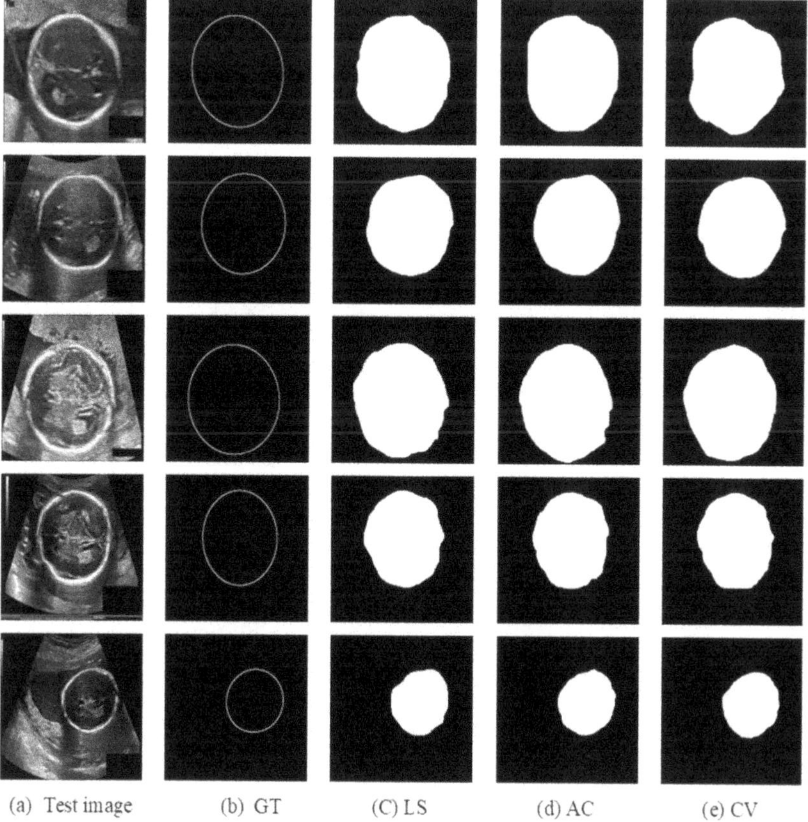

Figure 6.8. Results obtained with sample test images.

with Kapur's technique are better compared to Otsu's technique. The results obtained with the AC and the CV approaches are depicted in figures 6.10 and 6.11, respectively, and these measures are quite lesser compared with the LS technique.

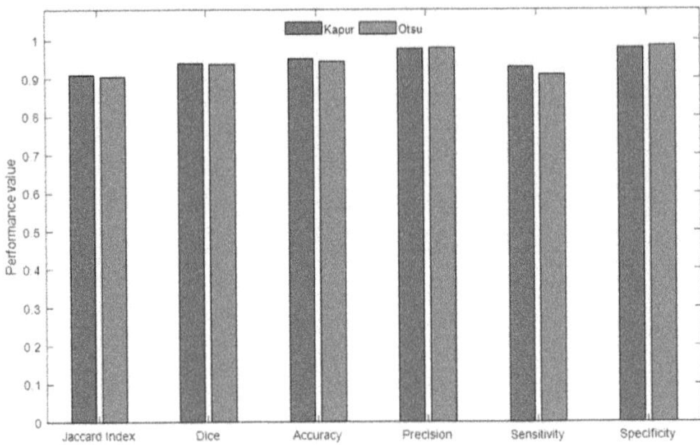

Figure 6.9. Assessment of the overall performance of the LS approach with Kapur's and Otsu's techniques.

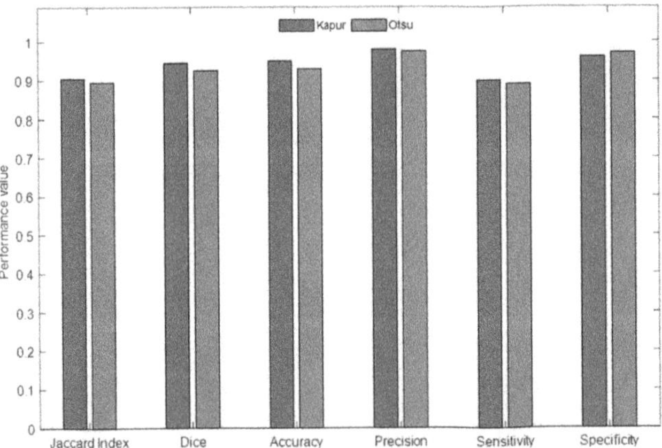

Figure 6.10. Assessment of the overall performance of the AC approach with Kapur's and Otsu's techniques.

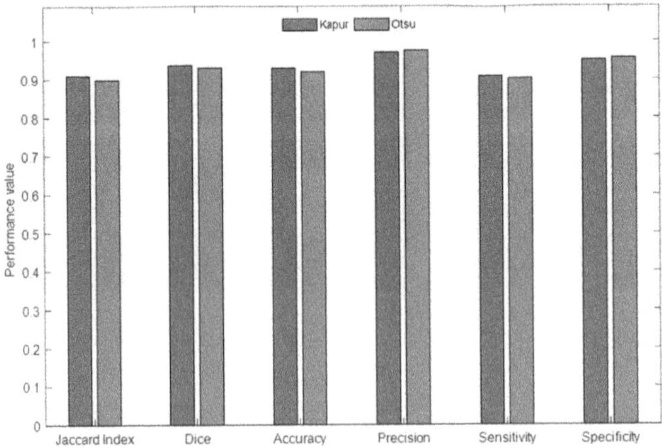

Figure 6.11. Assessment of the overall performance of the CV approach with Kapur's and Otsu's techniques.

From these results it can be confirmed that the proposed technique helps to extract the essential head section from the 2D ultrasound image and the extracted section may be considered to evaluate the FHC to assess the growth rate of the fetus with better accuracy. The JI, DC, and segmentation accuracy obtained with the LS approach can be improved by employing convolutional neural network (CNN) based image segmentation and assessment procedures. The considered HC18 dataset consists of 1334 clinical-grade ultrasound images and in the future all these images can be considered for the assessment and evaluation of the proposed computer-based FHC assessment system.

6.6 Conclusion

Prenatal viewing and image-based assessment of fetal growth are essential procedures to assess the various conditions of the fetus. If any abnormality is detected in its initial stage, then the doctor will take the necessary steps to solve the problem and help ensure a safe growing environment for the fetus. In the medical domain, assessment of fetal growth using 2D ultrasound images is a common procedure and in the literature a considerable number of image examination techniques have been proposed to monitor fetal growth. The proposed work implemented a computer-assisted image examination procedure with pre-processing and post-processing techniques and the outcome was then used to evaluate fetal growth. This work employed tri-level thresholding with the Kapur or Otsu techniques and the optimal threshold was identified with a chosen HA. This work implemented a comparative assessment between PS, BA, and FA and the experimental outcome confirmed that the PSO offered better optimization convergence than BA or FA. After the enhancement, the essential image section is then extracted using the LS, AC, or CV segmentation approach. The results observed in this proposed study confirmed that Kapur's thresholding combined with the LS helps to obtain better results compared to the AC and CV approaches. The proposed methodology also presented better values of the JI, DC, and accuracy. These values confirm that the proposed technique is clinically significant, and in the future it can be used to examine clinical-grade 2D ultrasound images.

References

[1] Sree S V, Ng E Y, Acharya R U and Faust O 2011 Breast imaging: a survey *World J. Clin. Oncol.* **2** 171

[2] Acharya U R, Tong J, Sree S V, Chua C K, Ha T P, GhistaDN, Chattopadhyay S, Ng K H and Suri J S 2012 Computer-based identification of type 2 diabetic subjects with and without neuropathy using dynamic plantar pressure and principal component analysis *J. Med. Syst.* **36** 2483–91

[3] Swapna G, Ghista D N, Martis R J, Ang A P and Sree S V 2012 ECG signal generation and heart rate variability signal extraction: signal processing, features detection, and their correlation with cardiac diseases *J. Mech. Med. Biol.* **12** 1240012

[4] Acharya U R, Chua E C, Chua K C, Min L C and Tamura T 2010 Analysis and automatic identification of sleep stages using higher-order spectra *Int. J. Neural Syst.* **20** 509–21

[5] Giri D, Acharya U R, Martis R J, Sree S V, Lim T C, VI T A and Suri J S 2013 Automated diagnosis of coronary artery disease affected patients using LDA, PCA, ICA and discrete wavelet transform *Knowl.-Based Syst.* **37** 274–82

[6] Acharya U R, Ng E Y, Tan J H, Sree S V and Ng K H 2012 An integrated index for the identification of diabetic retinopathy stages using texture parameters *J. Med. Syst.* **36** 2011–20

[7] Acharya U R, Sree S V, Chattopadhyay S and Suri J S 2012 Automated diagnosis of normal and alcoholic EEG signals *Int. J. Neural Syst.* **22** 1250011

[8] Acharya U R, Sree S V, Krishnan M M, Molinari F, Garberoglio R and Suri J S 2012 Non-invasive automated 3D thyroid lesion classification in ultrasound: a class of ThyroScan™ systems *Ultrasonics* **52** 508–20

[9] Acharya U R, Mookiah M R, Sree S V, Afonso D, Sanches J, Shafique S, Nicolaides A, Pedro L M, Fernandes J F and Suri J S 2013 Atherosclerotic plaque tissue characterization in 2D ultrasound longitudinal carotid scans for automated classification: a paradigm for stroke risk assessment *Med. Biol. Eng. Comput.* **51** 513–23

[10] Acharya U R, Fernandes S L, WeiKoh J E, Ciaccio E J, Fabell M K, Tanik U J, Rajinikanth V and Yeong C H 2019 Automated detection of Alzheimer's disease using brain MRI images—a study with various feature extraction techniques *J. Med. Syst.* **43** 302

[11] Jahmunah V, Oh S L, Rajinikanth V, CiaccioEJ, Cheong K H, Arunkumar N and Acharya U R 2019 Automated detection of schizophrenia using nonlinear signal processing methods *Artif. Intell. Med.* **100** 101698

[12] Kuppermann M, Norton M E, Gates E, Gregorich S E, Learman L A, Nakagawa S, Feldstein V A, Lewis J, Washington A E and Nease R F 2009 Computerized prenatal genetic testing decision-assisting tool: a randomized controlled trial *Obstet. Gynecol.* **113** 53–63

[13] Merz E 1999 3-D ultrasound in prenatal diagnosis *Curr. Obstet. Gynecol.* **9** 93–100

[14] DiPietro J A, Ghera M M and Costigan K A 2008 Prenatal origins of temperamental reactivity in early infancy *Early Hum. Dev.* **84** 569–75

[15] Hadlock F P, Harrist R B and Martinez-Poyer J 1991 *In utero* analysis of fetal growth: a sonographic weight standard *Radiology* **181** 129–33

[16] Hadlock F P, Harrist R B, Carpenter R J, Deter R L and Park S K 1984 Sonographic estimation of fetal weight. The value of femur length in addition to head and abdomen measurements *Radiology* **150** 535–40

[17] Yoshida S, Unno N, Kagawa H, Shinozuka N, Kozuma S and Taketani Y 2000 Prenatal detection of a high-risk group for intrauterine growth restriction based on sonographic fetal biometry *Int. J. Gynaecol. Obstet.* **68** 225–32

[18] Kalish R B and Chervenak F A 2005 Sonographic determination of gestational age *Ultrasound Review Obstet. Gynecol.* **5** 254–8

[19] Van den Heuvel T L, de Bruijn D, de Korte C L and Ginneken B V 2018 Automated measurement of fetal head circumference using 2D ultrasound images *PLoS One* **13** e0200412

[20] Thanaraj R I, Anand B, Rahul J A and Rajinikanth V 2020 Appraisal of breast ultrasound image using Shannon's thresholding and level-set segmentation *Progress in Computing, Analytics and Networking* (Singapore: Springer) pp 621–30

[21] Hadlock F P, Deter R L, Harrist R B and Park S K 1982 Fetal head circumference: relation to menstrual age *Am. J. Roentgenol.* **138** 649–53

[22] Valsky D V, Lipschuetz M, Bord A, Eldar I, Messing B, Hochner-Celnikier D, Lavy Y, Cohen S M and Yagel S 2009 Fetal head circumference and length of second stage of labor are risk factors for levatorani muscle injury, diagnosed by 3-dimensional transperineal ultrasound in primiparous women *Am. J. Obstet. Gynecol.* **201** 91–e1

[23] Elvander C, Högberg U and Ekeus C 2012 The influence of fetal head circumference on labor outcome: a population-based register study *Acta Obstet. Gynecol. Scand.* **91** 470–5

[24] Lipschuetz M *et al* 2018 Sonographic large fetal head circumference and risk of cesarean delivery *Am. J. Obstet. Gynecol.* **218** 339.e1–7

[25] Li J, Wang Y, Lei B, Cheng J Z, Qin J, Wang T, Li S and Ni D 2017 Automatic fetal head circumference measurement in ultrasound using random forest and fast ellipse fitting *IEEE J. Biomed Health Inform.* **22** 215–23

[26] Sobhaninia Z, Rafiei S, Emami A, Karimi N, Najarian K, Samavi S and Soroushmehr S R 2019 Fetal ultrasound image segmentation for measuring biometric parameters using multi-task deep learning *2019 41st Annual Int. Conf. of the IEEE Engineering in Medicine and Biology Society (EMBC)* (Piscataway, NJ: IEEE) pp 6545–8

[27] Rajinikanth V, Dey N, Kumar R, Panneerselvam J and Raja N S 2019 Fetal head periphery extraction from ultrasound image using Jaya algorithm and Chan–Vese segmentation *Procedia Comput. Sci.* **152** 66–73

[28] Yushkevich P A, Gao Y and Gerig G 2016 ITK-SNAP: an interactive tool for semi-automatic segmentation of multi-modality biomedical images *38th Annual Int. Conf. of the IEEE Engineering in Medicine and Biology Society (EMBC)* (Piscataway, NJ: IEEE) pp 3342–5

[29] Bhandary A, Prabhu G A, Rajinikanth V, ThanarajKP, Satapathy S C, Robbins D E, Shasky C, Zhang Y D, Tavares J M and Raja N S 2020 Deep-learning framework to detect lung abnormality—a study with chest x-ray and lung CT scan images *Pattern Recognit. Lett.* **129** 271–8

[30] Fernandes S L, Tanik U J, Rajinikanth V and Karthik K A 2020 A reliable framework for accurate brain image examination and treatment planning based on early diagnosis support for clinicians *Neural. Comput. Appl.* **32** 15897–908

[31] Raja N, Rajinikanth V, Fernandes S L and Satapathy S C 2017 Segmentation of breast thermal images using Kapur's entropy and hidden Markov random field *J. Med. Imaging Health Inform.* **7** 1825–9

[32] Rajinikanth V, Madhavaraja N, Satapathy S C and Fernandes S L 2017 Otsu's multi-thresholding and active contour snake model to segment dermoscopy images *J. Med. Imaging Health Inform.* **7** 1837–40

[33] Rajinikanth V, Fernandes S L, Bhushan B and Sunder N R 2018 Segmentation and analysis of brain tumor using Tsallis entropy and regularised level set *Proc. of 2nd Int. Conference on Micro-Electronics, Electromagnetics and Telecommunications* (Singapore: Springer) pp 313–21

[34] Kennedy J and Eberhart R 1995 Particle swarm optimization *Proc. of ICNN'95-Int. Conf. on Neural Networks* vol 4 (Piscataway, NJ: IEEE) pp 1942–8

[35] Rajinikanth V, Raja N S and Satapathy S C 2016 Robust color image multi-thresholding using between-class variance and cuckoo search algorithm *Information Systems Design and Intelligent Applications* (New Delhi: Springer) pp 379–86

[36] Yang X S 2011 Bat algorithm for multi-objective optimisation *Int. J. Bio-Inspired Comput.* **3** 267–74

[37] Dey N and Rajinikanth V 2021 Applications of Bat Algorithm and its Variants (Singapore: Springer)

[38] Rajinikanth V and Couceiro M S 2015 Optimal multilevel image threshold selection using a novel objective function *Information Systems Design and Intelligent Applications* (New Delhi: Springer) pp 177–86

[39] Yang X S 2010 Firefly algorithm, stochastic test functions and design optimisation *Int. J. Bio-Inspired Comput.* **2** 78–84

[40] Yang X S and He X 2013 Firefly algorithm: recent advances and applications *Int. J Swarm Intell.* **1** 36–50

[41] Rajinikanth V, Raja N S and Kamalanand K 2017 Firefly algorithm assisted segmentation of tumor from brain MRI using Tsallis function and Markov random field *J. Control. Eng. Appl. Inform* **19** 97–106

[42] Sri Madhava Raja N, Rajinikanth V and Latha K 2014 Otsu based optimal multilevel image thresholding using firefly algorithm *Model. Simul. Eng.* **2014** 794574

[43] Rajinikanth V, Satapathy S C, Dey N, Fernandes S L and Manic K S 2019 Skin melanoma assessment using Kapur's entropy and level set—a study with Bat algorithm *Smart Intelligent Computing and Applications* (Singapore: Springer) pp 193–202

[44] Kapur J N, Sahoo P K and Wong A K 1985 A new method for gray-level picture thresholding using the entropy of the histogram *Comput. Vis. Graph. Image Process.* **29** 273–85

[45] Kumar U, Kumar V and Kapur J N 1986 Normalized measures of entropy *Int. J. General System.* **12** 55–69

[46] Kapur J N and Kesavan H K 1992 Entropy optimization principles and their applications *Entropy and Energy Dissipation in Water Resources* (Dordrecht: Springer) pp 3–20

[47] Shree T V, Revanth K, Raja N S and Rajinikanth V 2018 A hybrid image processing approach to examine abnormality in retinal optic disc *Proc. Comp. Sci.* **125** 157–64

[48] Rajinikanth V, Thanaraj K P, Satapathy S C, Fernandes S L and Dey N 2019 Shannon's entropy and watershed algorithm based technique to inspect ischemic stroke wound *Smart Intelligent Computing and Applications* (Singapore: Springer) pp 23–31

[49] Otsu N 1979 A threshold selection method from gray-level histograms *IEEE Trans. Syst. Man Cybern.* **9** 62–6

[50] Rajinikanth V, Raja N S and Satapathy S C 2016 Robust color image multi-thresholding using between-class variance and cuckoo search algorithm *Information Systems Design and Intelligent Applications* (New Delhi: Springer) pp 379–86

[51] Li C, Xu C, Gui C and Fox M D 2010 Distance regularized level set evolution and its application to image segmentation *IEEE Trans. Image Process.* **19** 3243–54

[52] Kass M, Witkin A and Terzopoulos D 1988 Snakes: active contour models *Int. J. Comput. Vision* **1** 321–31

[53] Vese L A and Chan T F 2002 A multiphase level set framework for image segmentation using the Mumford and Shah model *Int. J. Comput. Vision* **50** 271–93

[54] Getreuer P 2012 Chan–Vese segmentation *Image Process. On Line* **2** 214–24

[55] Rajinikanth V, Satapathy S C, Fernandes S L and Nachiappan S 2017 Entropy based segmentation of tumor from brain MR images—a study with teaching learning based optimization *Pattern Recognit. Lett.* **94** 87–95

[56] Rajinikanth V and Satapathy S C 2018 Segmentation of ischemic stroke lesion in brain MRI based on social group optimization and fuzzy-Tsallis entropy *Arab. J. Sci. Eng.* **43** 4365–78

[57] Pugalenthi R, Rajakumar M P, Ramya J and Rajinikanth V 2019 Evaluation and classification of the brain tumor MRI using machine learning technique *J. Control Eng. Appl. Inform.* **21** 12–21

[58] Bakiya A, Kamalanand K, Rajinikanth V, Nayak R S and Kadry S 2020 Deep neural network assisted diagnosis of time-frequency transformed electromyograms *Multimedia Tools Appl.* **79** 11051–67

[59] Vinnarasi S F, Rose J A and Rajinikanth V 2020 An approach to extract low-grade tumor from brain MRI slice using soft-computing scheme *Progress in Computing, Analytics and Networking* (Singapore: Springer) pp 273–82

[60] Rose J A, Vinnarasi S F and Rajinikanth V 2020 Assessment of fundus images for retinal abnormality screening—a study *Progress in Computing, Analytics and Networking* (Singapore: Springer) pp 303–12

IOP Publishing

Predictive Analytics in Healthcare, Volume 1
Transforming the future of medicine
Vinithasree Subbhuraam

Chapter 7

Classification of retinal fundus images into healthy/AMD classes using mayfly algorithm selected features

David Taniar, Seifedine Kadry and Venkatesan Rajinikanth

The eye is one of the vital sensory organs in human physiology and abnormalities in the eye can arise due to various factors including ageing. Age-related macular degeneration (AMD) is one of the common eye abnormalities in elderly people and untreated AMD leads to loss of vision. This research aims to develop a machine-learning (ML) procedure to evaluate and categorize fundus images into healthy or AMD classes. This framework consists of the following phases: (i) image selection and pre-processing, (ii) feature extraction, (iii) feature reduction using the mayfly optimization algorithm (MOA), and (iv) binary classification and validation. In this work, benchmark fundus images are considered for investigation. The essential image features are mined using the gray level co-occurrence matrix (GLCM), local binary pattern (LBP), and entropies. The dominant features are then selected using the MOA and these features are then considered to train, test, and validate the performance of the classifiers considered in this research. In this work, a ten-fold cross validation is implemented and the best value is chosen as the final result. The experimental investigation of this research confirms that the ML scheme with the fine-tree classifier offers 94% accuracy compared to the other approaches.

7.1 Introduction

The eye is one of the primary sensory organs in human physiology, and is responsible for converting incident light into visual information. Abnormalities in the eye will severely affect the sensory system and hence vision related abnormalities must be treated with care [1–5].

In humans, abnormalities in the eye arise for a variety of reasons, such as birth defects, accidents, and eye problems due to ageing. Most eye abnormalities due to birth defects cannot be treated and accidental abnormalities can be treated with

doi:10.1088/978-0-7503-2312-3ch7

recommended procedures. The number of above cases is much smaller compared to disease due to ageing and, hence, in recent years a considerable number of measures have been taken to reduce the effect of these diseases. In elderly people (age > 60 years) the eye abnormality called age -related macular degeneration (AMD) is commonly found irrespective of gender, and untreated AMD will lead to vision loss. The recent statement by the World Heath Organisation (WHO) keeps this abnormality in the priority eye diseases (PED) group [6]. This statement also confirms that AMD is the third most common cause of blindness in humans. The primary cause of AMD is the growth of a degenerative wound in the macula, a primary part of the retina. The wound will limit the functions of the retina by disturbing the blood flow to the macula [7–9].

The AMD is commonly categorized as dry or wet and clinical-level detection can be achieved with angiography, optical coherence tomography (OCT) and fundus retinal imaging (FI). Compared to the other two imaging modalities, FI will help to obtain clear information about the affected area. Hence, recently a number of AMD detection methods based on FI have been proposed and implemented in the literature [3, 10–12].

The aim of this research is to develop a significant machine-learning scheme (MLS) to detect AMD from FI. The essential test images considered in this research work are collected from the REFUGE 2018 dataset [13–15]. The considered dataset consists of 400 RGB scaled FI images collected using the clinical protocol (89 AMD and 311 normal images). To reduce the computational burden of the proposed MLS, every FI in this dataset is resized to 256 × 256 × 3 pixels. In this work 200 AMD images (89 original and 111 augmented) and 200 normal images are considered for the assessment.

The developed MLS consists of the following phases: (i) pre-processing to reduce the dimension of test image to 256 × 256 × 3 pixels, (ii) abnormal section extraction using saliency based enhancement and morphological segmentation, (iii) feature extraction using the chosen technique, (iv) MOA assisted feature reduction, and (v) classification of the images into normal and AMD classes.

In this work, the performance of the MLS is tested using a number of binary classifiers and the result obtained with the fine-tree classifier are superior (94%) compared to the classifiers considered in this research.

This chapter is organized as follows: section 7.2 presents the background of AMD detection, section 7.3 presents the methodology of this study, and sections 7.4 and 7.5 present the experimental outcome and conclusion, respectively.

7.2 AMD detection

Age-related diseases are very common in elderly people for a variety of reasons. AMD is one of the diseases found widely in elderly people and untreated AMD will lead to loss of vision.

Figure 7.1 shows FI of the normal and AMD classes and from these images it can be seen that AMD will cause abnormalities in the macula section of the eye and this will cause a severe eye problem.

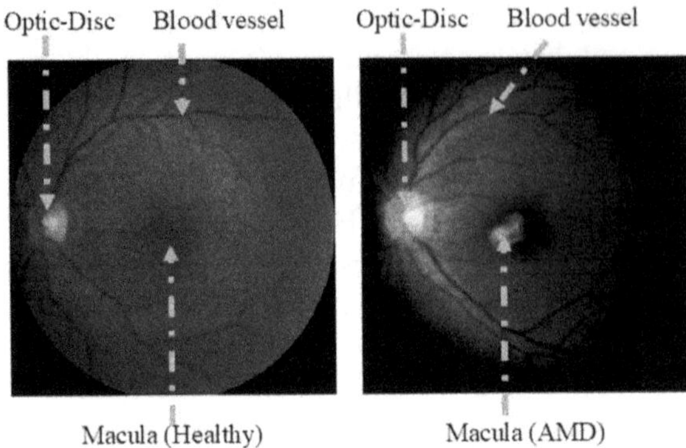

Figure 7.1. Sample retinal images of the healthy and AMD classes.

Recent work by Rajinikanth *et al* [3] proposed an MLS to categorize FI into normal and AMD classes and this work achieved a classification accuracy of 93.67% using a support vector machine (SVM) classifier. In this work, the essential image pre-processing was employed using thresholding as well as Gaussian filter based image enhancement. The fundamental features, such GLCM and entropies, are extracted from the pre-processed images and then the feature reduction is implemented with Student's *t*-test, which helps to obtain a feature sub-set of dimensions $1 \times 1 \times 19$. Finally, a binary classification is implemented and the performance of the classifier is confirmed using a ten-fold cross validation.

The aim of the present research is to implement a suitable methodology to improve the detection accuracy of the AMD. Along with the GLCM and entropies, this work additionally considers the LBP features to obtain improved accuracy in the AMD detection process.

7.3 Methodology

This section demonstrates the various stages of the proposed MLS. Figure 7.2 depicts the architecture of the proposed AMD recognition system.

Resizing is initially employed to reduce the image dimensions of the retinal test pictures. The converted image is then examined separately using: (i) tri-level thresholding based on MOA and Shannon's entropy (SE), (ii) Gaussian filter enhancement (GFE), and (iii) LBP. After the possible enhancement of the FI, the fundamental features are mined using the appropriate techniques, as depicted in the figure. Later, the dominant features are selected using the MOA with a chosen cost function and finally the performance of the proposed MLS is tested and validated using the binary classifiers.

7.3.1 Retinal image database

Image assisted eye abnormality assessment is an clinical procedure and is normally executed by an experienced ophthalmologist. In modern eye clinics, the abnormality assessment procedures are normally carried out using computerized algorithms and the

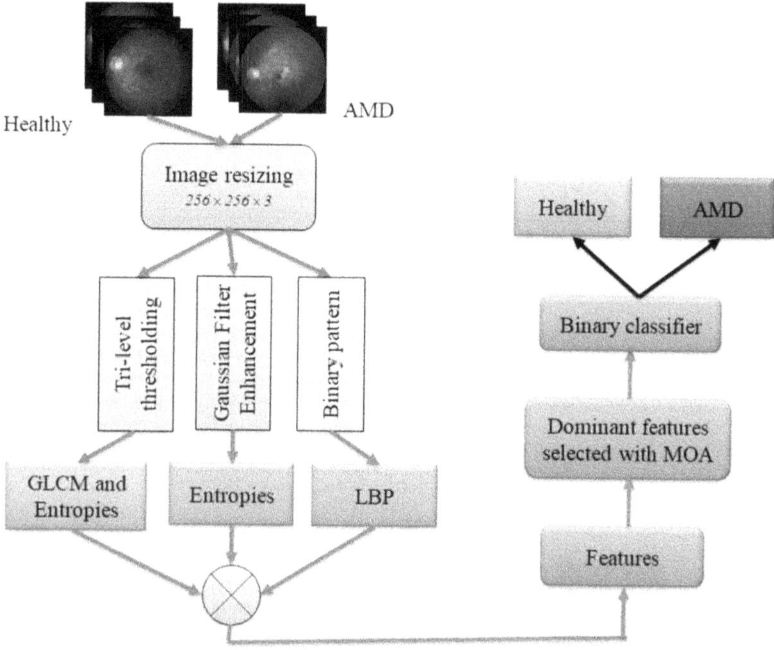

Figure 7.2. Architecture of the proposed AMD detection scheme.

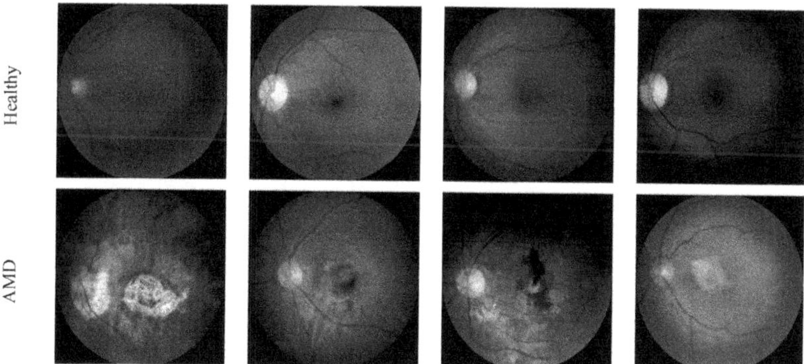

Figure 7.3. Sample test images of REFUGE's iChallenge-AMD.

outcome obtained with these computerized methodologies is then verified and confirmed by a doctor. Even though a considerable number of imaging modalities exist to record and evaluate retinal diseases, the FI is preferred by doctors due to its reliability. The proposed work considers the FI of the REFUGE 2018 dataset for the assessment. In this work, REFUGE's iChallenge-AMD images are considered for the investigation and from each class (healthy or AMD), 200 images are considered for the evaluation. The basic information regarding the test images can be found in [13–15].

Figure 7.3 depicts the sample test images of the chosen dataset and table 7.1 presents the information regarding the FI considered for the experimental

Table 7.1. The retinal images considered in this research.

Class	Dimension	Images		
		Total	Training	Testing
Healthy	256 × 256 × 3	200	150	50
AMD	256 × 256 × 3	200	150	50

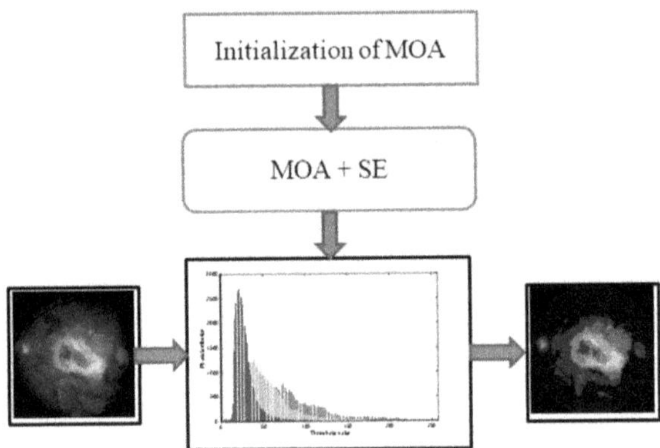

Figure 7.4. Thresholding supported image enhancement with MOA+SE.

investigation. In this work, 75% of the images are considered to train the binary classifiers and the remaining 25% of images are considered to test and validate the performance of the classifier. Based on the obtained classification accuracy, the performance of the developed MLS is confirmed.

7.3.2 Image thresholding

In the image processing domain, thresholding is adopted to enhance the chosen image region by selecting the optimal threshold. The considered FI is available in the RGB scale and the selection of the desired threshold is a challenging task. Hence, in this work, the MOA is adopted to identify the optimal threshold by enhancing the Shannon's entropy value. As discussed in earlier works [16–19], tri-level thresholding is considered to enhance the abnormal sections in the FI. Figure 7.4 depicts the tri-level thresholding employed with the MOA+SE on the RGB threshold of the AMD class FI.

During the tri-level threshold, the chosen approach is allowed to explore the R, G and B histogram of the image to find and group the pixels into three sections, as shown in figure 7.5. Figures 7.5(a) and (b) depict the image and histograms, respectively. In this work, the appropriate threshold is selected using the MOA and Shannon's entropy (MOA+SE).

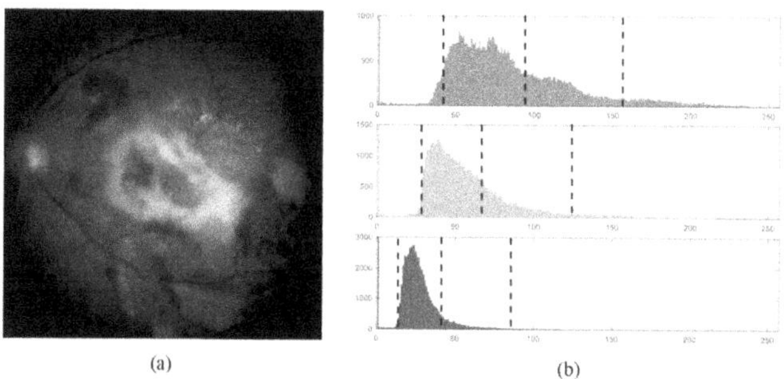

(a) (b)

Figure 7.5. RGB threshold of the chosen AMD class test picture.

7.3.2.1 The mayfly optimization algorithm

This is one of the recent nature-inspired techniques developed from the other well known algorithms, such as particle swarm optimization (PSO), firefly algorithm (FA), and genetic algorithm (GA) [20]. The phases of the MOA involve: (i) initialization of male and female flies, (ii) permitting the male mayfly to discover the G_{best} of the problem, (iii) permitting the female mayfly to identify and join with the male mayfly existing in G_{best}, (iv) offspring creation, and (v) stopping the search and delivering the optimal result [21, 22].

The search ability of MOA relies on the early location of the male and its distance towards the female. When every fly is randomly initiated in a chosen search location with the appropriate number of male and female agents, each fly is permitted to reach G_{best} according to the convergence. This process continues until an equal amount of offspring is obtained with every pair of agents. To terminate the search of the MOA at this stage, the weights for the offspring are assigned with a zero velocity. Basic details of the MOA can be found in [23, 24] and the code can be obtained from [25].

Figure 7.6(a) presents the initial position of the mayfly in a chosen search location. The number of male (X) and female (Y) flies are assigned with a chosen amount (N). Figure 7.6(b) presents the convergence of a male fly towards the optimal location and figure 7.6(c) depicts the final outcome of the MOA.

Let this approach consist of equal male (X) and female (Y) flies randomly initialized in a D-dimensional search space and every fly is depicted as follows: $i = 1, 2, \ldots, N$ (with a chosen value of $N = 30$). During this search, every agent is arbitrarily located in the chosen search space and each fly is permitted to converge towards the best location (G_{best}). As in figure 7.6(b), the male fly is permitted to converge to G_{best} by altering its location and speed. The convergence of the fly towards the optimal result will be monitored by the Cartesian distance (CD) with an increase in iterations. This process is similar to the movement of the fly in the FA. The position and velocity update expressions are

$$E_i^{t+1} = E_i^t + F_i^{t+1} \tag{7.1}$$

$$F_{i,j}^{t+1} = F_{i,j}^t + C_1 * e^{-\beta D_p{}^2}(P_{\text{best}_{i,j}} - E_{i,j}^t) + C_2 * e^{-\beta D_g{}^2}(G_{\text{best}_{i,j}} - E_{i,j}^t), \tag{7.2}$$

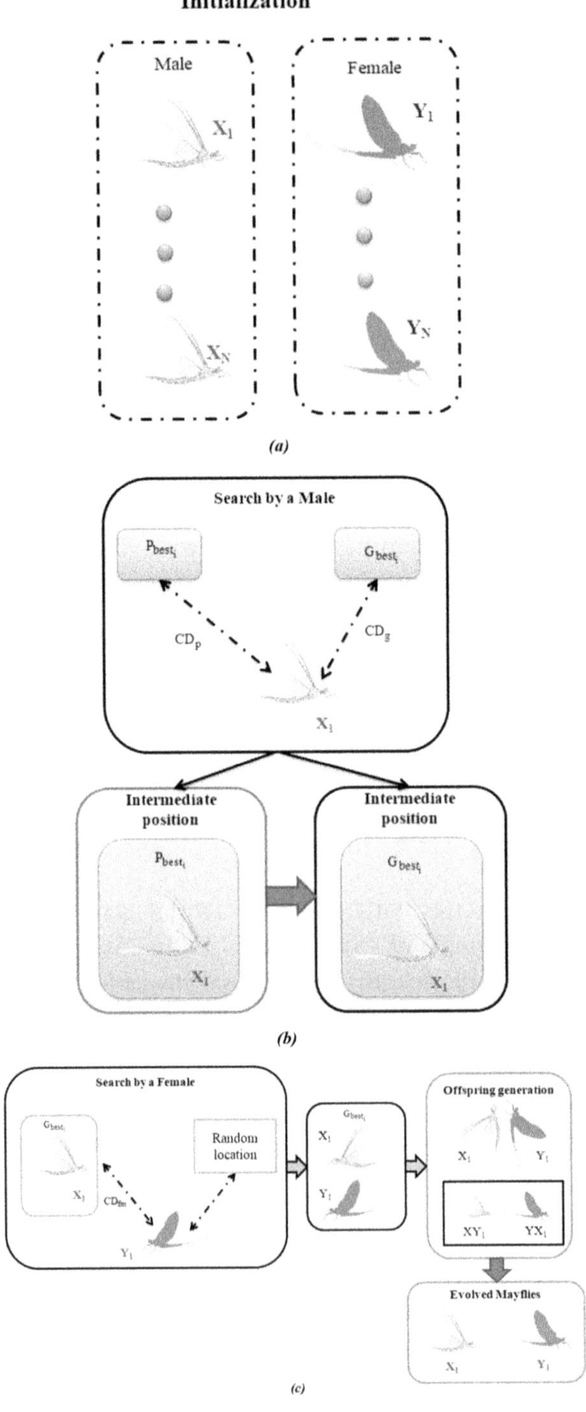

Figure 7.6. (a). Initialized mayfly in a chosen search space. (b) Intermediate position of the male mayfly. (c) Converged mayfly towards the optimal result.

where E_i^t and E_i^{t+1} are the first and converged positions, F_i^{t+1} and $F_{i,j}^{t+1}$ are the first and final velocities, respectively, the local learning parameter $(C_1) = 1$, the global learning parameter $(C_2) = 1.5$, $\beta = 2$, and D_p and D_g are the Cartesian distance. Equation (7.2) is constructed by integrating the FA and PSO. When the update of flies continues, every M will reach G_{best} and perform a velocity update to attract the F by performing a unique nuptial dance-like movement.

The velocity update during this process can be defined as

$$F_{i,j}^{t+1} = F_{i,j}^t + d^*R, \tag{7.3}$$

where the nuptial dance value $(d) = 5$ and R = random numeral $[-1, 1]$.

When the search by the M is completed, every F is then allowed to find an M converged at G_{best}. Figure 7.6(c) depicts the search by the F, which moves towards the M based on a distance (D_{mf}) or the Y escapes to a new section with a random walk $(W = 1)$.

The expression for position and velocity updates for the F is

$$E_i^{\prime t+1} = E_i^{\prime t} + F_i^{\prime t+1} \tag{7.4}$$

$$F_{i,j}^{\prime t+1} = \begin{cases} F_{i,j}^{\prime t} + C_2 e^{-\beta D_{mf}^2}(M_{i,j}^t - Y_{i,j}^t) & \text{if } O(F_i) > O(M_i) \\ F_{i,j}^{\prime t} + W^*r & \text{if } O(F_i) \leqslant O(M_i) \end{cases}, \tag{7.5}$$

where O is the objective function with maximal value.

When the iteration improves, every F will reach the M and the offspring generation takes place as depicted in figure 7.6(c). Other essential information on the MOA can be found in [20–24].

7.3.2.2 Shannon's entropy

Image pre-processing with SE is a commonly considered procedure and a number of earlier works confirm that SE helps to achieved a better result on a considerable number of medical images [26]. A detailed explanation of SE can be found in [16–18]. Previous work on SE implemented on AMD class FI can be found in [3]. The MOA combined with SE helps to obtain enhanced images with an assigned threshold of three and the gray version of this image is considered to obtain the GLCM and entropy features.

7.3.3 Gaussian filter enhancement

Gaussian filter based image enhancement improves the texture and edges of the image based on the chosen filter scale and in this work the GFE implemented in a recent work is adopted to pre-process the healthy and AMD class FI [27, 28].

7.3.4 Local binary pattern

LBP based image pre-processing is employed widely in the medical image examination domain and this enhancement can be implemented on the grayscale version of the FI. Previous works using the LBP can be found in [29–31] and the complete information about the LBP implemented in this work can be found in Gudigar *et al* [32]. In the proposed work, the LBP obtained for a chosen global weight of 4 is considered and this image helps to obtain an LBP feature of dimensions $1 \times 1 \times 59$.

7.3.5 Feature extraction

In the proposed work, the handcrafted features such as GLCM, entropies, and LBP are extracted from the pre-processed images. The entropy values, namely Kapur, Tsallis, Renyi, and Shannon [33–40], are extracted from the pre-processed FI. The GLCM features are also extracted from the thresholded images. The LBP features are extracted from the LBP enhanced images.

The features extracted are as follows:

$$\text{GLCM of threshold image} = 1 \times 1 \times 22 \tag{7.6}$$

$$\text{Entropies of threshold image} = 1 \times 1 \times 4 \tag{7.7}$$

$$\text{Entropies of GFE images} = (1 \times 1 \times 4) + ((1 \times 1 \times 4)) = 1 \times 1 \times 8 \tag{7.8}$$

$$\text{LBP} = 1 \times 1 \times 59 \tag{7.9}$$

$$\begin{aligned}\text{Total features} &= (1 \times 1 \times 22) + (1 \times 1 \times 4) + (1 \times 1 \times 8) + (1 \times 1 \times 59) \\ &= 1 \times 1 \times 93.\end{aligned} \tag{7.10}$$

7.3.6 Feature selection

The performance of a computerized disease detection system depends mainly on the features considered to categorize the test images into the healthy and the disease classes. The detection speed also depends on the features extracted from the test pictures. In order to avoid the overfitting problem, feature reduction procedures are normally adopted in class deep-learning and MLSs [41–46]. In the feature reduction process, a feature sub-set is selected using traditional and modern techniques. The traditional technique implements a chosen statistical method to find the dominant features between the healthy and disease classes. In the modern approach, heuristic algorithms are implemented to identify the possible feature sub-set to classify the test images.

In the proposed work, the MOA is implemented to identify the feature sub-set based on the CD between the features of the healthy and AMD classes. In this work, the MOA is allowed to search the features until the CD between the features are maximum. The implementation of a similar technique can be found in previous studies.

The MOA supported feature selection is depicted in figure 7.7 and this procedure helps to obtain the feature sub-set based on the maximized CD. This process helps to obtain a feature sub-set with dimensions $1 \times 1 \times 16$.

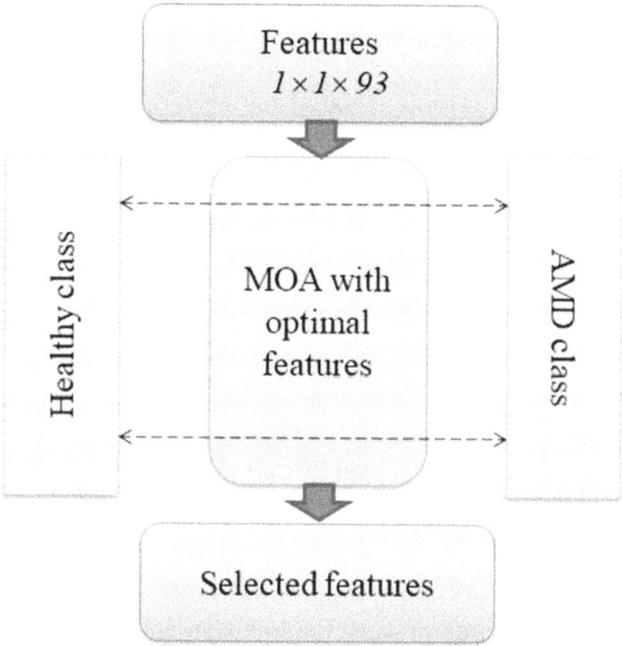

Figure 7.7. Feature selection with the MOA.

7.3.7 Classifier implementation and validation

The overall performance of the proposed MLS depends on the classifiers employed to categorize the images into the chosen classes. In this work, a binary classifier is employed to classify the FI into healthy and AMC classes. The classifiers decision tree (DT), SVM, and k-nearest neighbor (KNN) are considered in this work and complete information regarding these classifiers can be found in [47].

7.3.8 Performance measures

The eminence of an MLS relies on the performance obtained with the binary classifier. In this work the commonly used measures discussed in earlier works are considered [3].

The mathematical expression of the measures are as follows:

$$\text{Accuracy} = \text{AC} = \frac{\text{TP} + \text{TN}}{\text{TP} + \text{TN} + \text{FP} + \text{FN}} \tag{7.11}$$

$$\text{Precision} = \text{PR} = \frac{\text{TP}}{\text{TP} + \text{FP}} \tag{7.12}$$

$$\text{Sensitivity} = \text{SE} = \frac{\text{TP}}{\text{TP} + \text{FN}} \tag{7.13}$$

$$\text{Specificity} = \text{SP} = \frac{\text{TN}}{\text{TN} + \text{FP}} \tag{7.14}$$

$$\text{Negative predictive value} = \text{NP} = \frac{\text{TN}}{\text{TN} + \text{FN}} \tag{7.15}$$

$$F_1\text{-score} = F_1 = \frac{2\text{TP}}{2\text{TP} + \text{FN} + \text{FP}}, \tag{7.16}$$

where TP = true positive, TN = true negative, FP = false positive, and FN = false negative.

7.4 Results and discussion

This section presents the experimental results obtained in this research and the discussions thereof. The proposed experimental work considered the benchmark FI of the REFUGE 2018 dataset and this research work was implemented with Matlab software.

Initially, the tri-level thresholding is implemented on the chosen test images with MOA+SE. During this search process, the following initial values are assigned for the MOA: agent size (N) = 30, dimension = 3, iteration value (M_{Iter}) = 2500, objective function= maximized SE, and terminating value = M_{Iter} or maximized SE.

Every FI in the chosen dataset is pre-processed using the MOA+SE supported tri-level thresholding and from these images the features such as GLCM and entropies are extracted. After collecting the essential features from the thresholded images, the considered test images are further enhanced using other chosen procedures such as GFE and LBP and the essential features from these images are also collected.

Figure 7.8 depicts the sample FI and the pre-processed images. Figure 7.8(a) presents the test FI, figures 7.8(b)–(d) show the pre-processed images with thresholding, GFE, and LBP, respectively. Every technique improved the texture

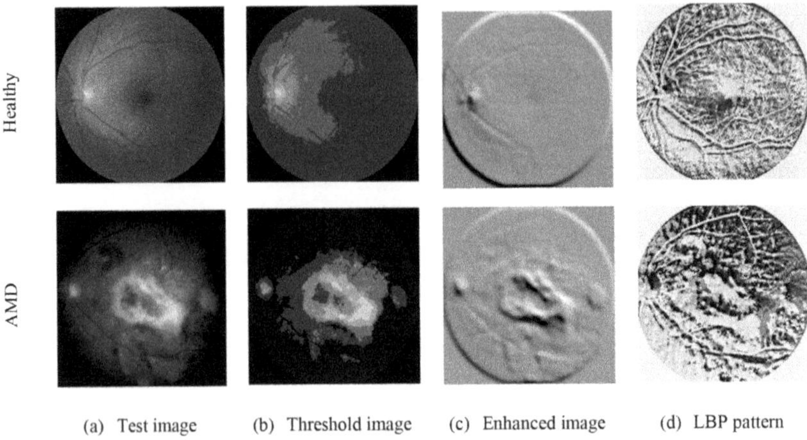

Figure 7.8. Sample test images and the pre-processed image.

information of the FI considerably and the enhanced texture information is extracted as the vital features. After extracting the necessary features and selecting the dominant feature sub-set, binary classifiers are implemented and validated.

In the proposed work, all the features (1 × 1 × 16) selected by the MOA are considered to train and validate the considered classifiers, namely DT, SVM, and KNN and their variants. In this work, the features collected from 150 FI are considered to train the classifiers and the features of 50 FI are considered to validate the classifier performance. During this process, a ten-fold cross validation is considered and the best result of the trial is chosen as the final result.

The classifier result obtained with the considered MLS with various binary classifiers is presented in table 7.2. From this table, it can be noted that the accuracy obtained with $DT_{FineTree}$ is better (94%) compared with other classifiers adopted in this research. The overall performances of the classifiers are also verified using the glyph plot as depicted in figure 7.9. The pattern which occupies a larger area is

Table 7.2. Disease detection performance of the chosen binary classifiers.

Classifier	TP	FN	TN	FP	AC	PR	SE	SP	NP	F_1
$DT_{CoarseTree}$	44	6	41	9	85.00	83.02	88.00	82.00	87.23	85.44
$DT_{MediumTree}$	46	4	47	3	93.00	93.88	92.00	94.00	92.16	92.93
$DT_{FineTree}$	47	3	47	3	94.00	94.00	94.00	94.00	94.00	94.00
SVM_{Linear}	45	5	46	4	91.00	91.84	90.00	92.00	90.20	90.91
SVM_{RBF}	44	6	45	5	89.00	89.79	88.00	90.00	88.23	88.89
KNN_{Coarse}	41	9	42	8	83.00	83.67	82.00	84.00	82.35	82.83
KNN_{Medium}	47	3	44	6	91.00	88.68	94.00	88.00	93.62	91.26
KNN_{Fine}	46	4	46	4	92.00	92.00	92.00	92.00	92.00	92.00

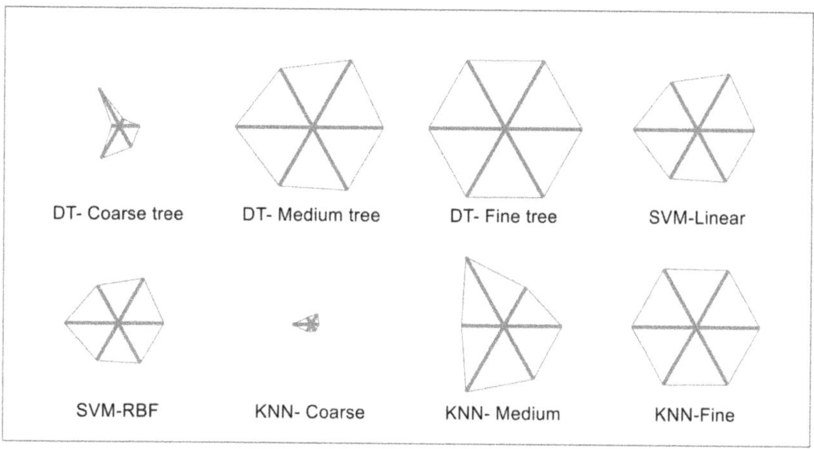

Figure 7.9. Glyph plot of the performance values of the binary classifiers.

considered as superior and the result of fine tree is is superior to the other variants of DT, SVM, and KNN.

Table 7.2 confirms that the fine tree of DT performs better compared to other methods. To confirm the overall performance, the glyph plot constructed using all the size measures confirms that the result using DT and KNN with coarse provided smaller measures compared to the other methods, such as the medium and fine categories. The result of the SVM is also very poor compared to the fine tree classifier. To validate the performance of the proposed MLS, the result obtained with the earlier research is considered [3]. The performance of the classifiers naïve-Bayes (NB), DT, KNN, random forest (RF), and SVM, discussed in a similar previous study [3] is compared with the results of the proposed work and a graphical representation of the comparison is depicted in figure 7.10.

The result of the proposed work confirms that the classification accuracy of the proposed scheme is superior compared to the previous work. This improvement is due to the implementation of the LBP and identification of the dominant feature sub-set with the MOA. In future, the performance of the MLS can be improved by considering other handcrafted features existing in the medical image examination literature.

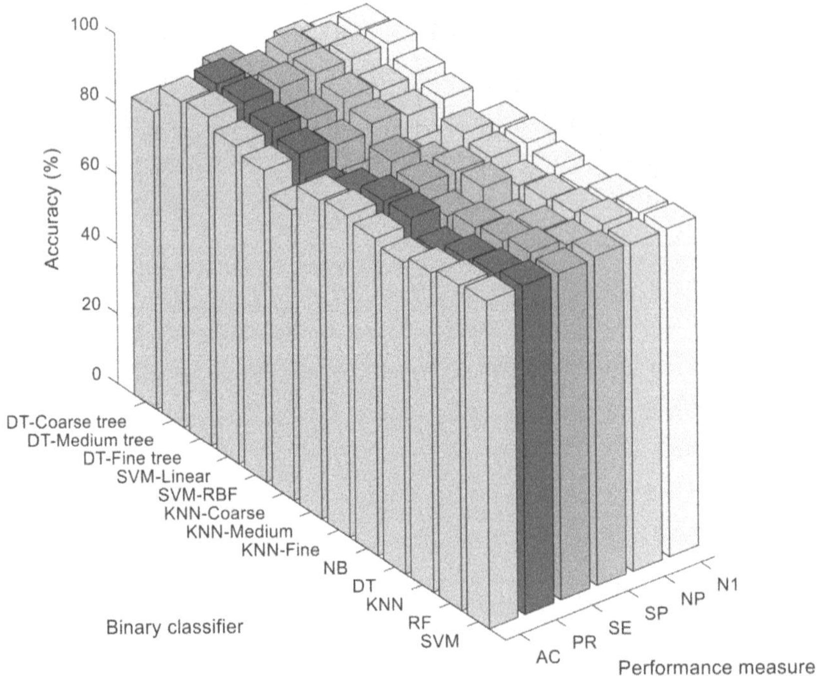

Figure 7.10. Comparison of the obtained results with previous work.

7.5 Conclusion

Automated disease detection is a necessary procedure to reduce the computational burden faced during mass screening processes. In the proposed research, an MLS is developed to support the automated classification of FI into healthy and AMD classes. The various phases involved in the developed MLS are (i) FI collection and resizing, (ii) pre-processing of FI with tri-level thresholding, GFE, and LBP, (iii) extraction of GLCM, entropy, and LBP features, (iv) MOA based feature reduction, and (v) classification and validation. In this work $1 \times 1 \times 16$ features are considered for the classification and a ten-fold cross validation is also employed. The classification accuracy obtained using fine tree is (94%), which is higher than that of the other classifiers in this research. The results of this research confirm that the proposed MLS helps to obtain better results on the REFUGE 2018 dataset compared to previous research.

References

[1] Koh J E *et al* 2017 Diagnosis of retinal health in digital fundus images using continuous wavelet transform (CWT) and entropies *Comput. Biol. Med.* **84** 89–97

[2] Shree T V, Revanth K, Raja N S and Rajinikanth V 2018 A hybrid image processing approach to examine abnormality in retinal optic disc *Procedia Comput. Sci.* **125** 157–64

[3] Rajinikanth V, Sivakumar R, Hemanth D J, Kadry S, Mohanty J R, Arunmozhi S, Raja N S and Nhu N G 2021 Automated classification of retinal images into AMD/non-AMD class —a study using multi-threshold and Gaussian-filter enhanced images *Evol. Intell.* **14** 1163–71

[4] Rajinikanth V, Lin H, Panneerselvam J and Raja N S 2020 Examination of retinal anatomical structures—a study with spider monkey optimization algorithm *Applied Nature-Inspired Computing: Algorithms and Case Studies* (Singapore: Springer) pp 177–97

[5] Rose J A, Vinnarasi S F and Rajinikanth V 2020 Assessment of fundus images for retinal abnormality screening—a study *Progress in Computing, Analytics and Networking* (Singapore: Springer) pp 303–12

[6] WHO 2011 Prevention of blindness and visual impairment. The distribution of blind and visually impaired of all ages in the six WHO regions https://www.who.int/blindness/GLOBALDATAFINALforweb.pdf

[7] Lim L S, Mitchell P, Seddon J M, Holz F G and Wong T Y 2012 Age-related macular degeneration *Lancet* **379** 1728–38

[8] Jager R D, Mieler W F and Miller J W 2008 Age-related macular degeneration *New Engl. J. Med.* **358** 2606–17

[9] Bressler N M, Bressler S B and Fine S L 1988 Age-related macular degeneration *Survey Ophthalmol.* **32** 375–413

[10] Corvi F, Cozzi M, Invernizzi A, Pace L, Sadda S R and Staurenghi G 2021 Optical coherence tomography angiography for detection of macular neovascularization associated with atrophy in age-related macular degeneration *Graefe's Arch. Clin. Exp. Ophthalmol.* **259** 291–9

[11] Ho A C *et al* 2021 Real-world performance of a self-operated home monitoring system for early detection of neovascular age-related macular degeneration *J. Clin. Med.* **10** 1355

[12] Ward E *et al* 2021 Monitoring for neovascular age-related macular degeneration (AMD) reactivation at home: the MONARCH study *Eye* **35** 592–600

[13] Li X, Jia M, Islam M T, Yu L and Xing L 2020 Self-supervised feature learning via exploiting multi-modal data for retinal disease diagnosis *IEEE Trans. Med. Imaging* **39** 4023–33
[14] Sun T and Oruc I A TeleAEye: Low-Cost Automated Eye Disease Diagnosis Using a Novel Smartphone Fundus Camera With AI https://dfantsun.com/scireports/GVRSF.pdf
[15] Rajinikanth V, Thanaraj K P, Satapathy S C, Fernandes S L and Dey N 2019 Shannon's entropy and watershed algorithm based technique to inspect ischemic stroke wound *Smart Intelligent Computing and Applications* (Singapore: Springer) pp 23–31
[16] Monisha R, Mrinalini R, Britto M N, Ramakrishnan R and Rajinikanth V 2019 Social group optimization and Shannon's function-based RGB image multi-level thresholding *Smart Intelligent Computing and Applications* (Singapore: Springer) pp 123–32
[17] Dey N, Rajinikanth V, Ashour A S and Tavares J M 2018 Social group optimization supported segmentation and evaluation of skin melanoma images *Symmetry* **10** 51
[18] Raja N S, Arunmozhi S, Lin H, Dey N and Rajinikanth V 2019 A study on segmentation of leukocyte image with Shannon's entropy *Histopathological Image Analysis in Medical Decision Making* (Hershey, PA: IGI Global) pp 1–27
[19] Zervoudakis K and Tsafarakis S 2020 A mayfly optimization algorithm *Comput. Ind. Eng.* **145** 106559
[20] Liu Z, Jiang P, Wang J and Zhang L 2021 Ensemble forecasting system for short-term wind speed forecasting based on optimal sub-model selection and multi-objective version of mayfly optimization algorithm *Expert Syst. Appl.* **177** 114974
[21] Abd Elaziz M, Senthilraja S, Zayed M E, Elsheikh A H, Mostafa R R and Lu S 2021 A new random vector functional link integrated with mayfly optimization algorithm for performance prediction of solar photovoltaic thermal collector combined with electrolytic hydrogen production system *Appl. Therm. Eng.* **193** 117055
[22] Juan Z H and Zheng-Ming G A 2020 Bare bones mayfly optimization algorithm *In* 2020 2nd Int. Conf. on Machine Learning, Big Data and Business Intelligence (MLBDBI) (Piscataway, NJ: IEEE) pp 238–41
[23] Gao Z M, Zhao J, Li S R and Hu Y R 2020 The improved mayfly optimization algorithm *J. Phys.: Conf. Series* **1684** 012077
[24] Zervoudakis K and Tsafarakis S 2020 Mayfly optimization algorithm—Matlab code v1 *Mendeley Data* **14** 106559
[25] Kannappan P L 1972 On Shannon's entropy, directed divergence and inaccuracy *Z. Wahrscheinlichkeit.* **22** 95–100
[26] Basu M 2002 Gaussian-based edge-detection methods—a survey *IEEE Trans. Syst. Man Cybern. Syst* C **32** 252–60
[27] Marr D and Hildreth E 1980 Theory of edge detection *Proc. R Soc. Lond.* B **207** 187–217
[28] Wang X, Han T X and Yan S 2009 An HOG-LBP human detector with partial occlusion handling *12th Int. Conf. on Computer Vision* (Piscataway, NJ: IEEE) pp 32–9
[29] Rao T and Kumar K S 2019 Retinal disease detection and classification using improved LBP technique *Proc. Int. Conf. on Sustainable Computing in Science, Technology and Management (SUSCOM)* (Jaipur: Amity University Rajasthan)
[30] Lemaitre G, Rastgoo M, Massich J, Sankar S, Mériaudeau F and Sidibé D 2015 Classification of SD-OCT volumes with LBP: application to DME detection *Ophthalmic Medical Image Analysis Int. Workshop 2*
[31] Gudigar A *et al* 2019 Global weighted LBP based entropy features for the assessment of pulmonary hypertension *Pattern Recognit. Lett.* **125** 35–41

[32] Sree S V, Ng E Y, Acharya R U and Faust O 2011 Breast imaging: a survey *World J. Clin. Oncol.* **2** 171

[33] Acharya U R, Tong J, Subbhuraam V S, Chua C K, Ha T P, Ghista D N, Chattopadhyay S, Ng K H and Suri J S 2012 Computer-based identification of type 2 diabetic subjects with and without neuropathy using dynamic planter pressure and principal component analysis *J. Med. Syst.* **36** 2483–91

[34] Swapna G, Ghista D N, Martis R J, Ang A P and Sree S V 2012 ECG signal generation and heart rate variability signal extraction: signal processing, features detection, and their correlation with cardiac diseases *J. Mech. Med. Biol.* **12** 1240012

[35] Acharya U R, Chua E C, Chua K C, Min L C and Tamura T 2010 Analysis and automatic identification of sleep stages using higher order spectra *Int. J. Neural Syst.* **20** 509–21

[36] Giri D, Acharya U R, Martis R J, Sree S V, Lim T C, VI T A and Suri J S 2013 Automated diagnosis of coronary artery disease affected patients using LDA, PCA, ICA and discrete wavelet transform *Knowl.-Based Syst.* **37** 274–82

[37] Acharya U R, Ng E Y, Tan J H, Sree S V and Ng K H 2012 An integrated index for the identification of diabetic retinopathy stages using texture parameters *J. Med. Syst.* **36** 2011–20

[38] Acharya U R, Sree S V, Chattopadhyay S and Suri J S 2012 Automated diagnosis of normal and alcoholic EEG signals *Int. J. Neural Syst.* **22** 1250011

[39] Acharya U R, Sree S V, Krishnan M M, Molinari F, Garberoglio R and Suri J S 2012 Non-invasive automated 3D thyroid lesion classification in ultrasound: a class of ThyroScan™ systems *Ultrasonics* **52** 508–20

[40] Rajinikanth V, Priya E, Lin H and Lin F 2021 *Hybrid Image Processing Methods for Medical Image Examination* (Boca Raton, FL: CRC Press)

[41] Bhandary A, Prabhu G A, Rajinikanth V, Thanaraj K P, Satapathy S C, Robbins D E, Shasky C, Zhang Y D, Tavares J M and Raja N S 2020 Deep-learning framework to detect lung abnormality–a study with chest x-ray and lung CT scan images *Pattern Recognit. Lett.* **129** 271–8

[42] Fernandes S L, Tanik U J, Rajinikanth V and Karthik K A 2020 A reliable framework for accurate brain image examination and treatment planning based on early diagnosis support for clinicians *Neural. Comput. Appl.* **32** 15897–908

[43] Raja N, Rajinikanth V, Fernandes S L and Satapathy S C 2017 Segmentation of breast thermal images using Kapur's entropy and hidden Markov random field *J. Med. Imaging Health Inform.* **7** 1825–9

[44] Rajinikanth V, Madhavaraja N, Satapathy S C and Fernandes S L 2017 Otsu's multi-thresholding and active contour snake model to segment dermoscopy images *J. Med. Imaging Health Inform.* **7** 1837–40

[45] Rajinikanth V and Kadry S 2021 Development of a framework for preserving the disease-evidence-information to support efficient disease diagnosis *Int. J. Data Warehous. Min.* **17** 63–84

[46] Khan M A, Kadry S, Alhaisoni M, Nam Y, Zhang Y, Rajinikanth V and Sarfraz M S 2020 Computer-aided gastrointestinal diseases analysis from wireless capsule endoscopy: a framework of best features selection *IEEE Access.* **8** 132850–9

[47] Raghavendra U, Gudigar A, Rao T N, Rajinikanth V, Ciaccio E J, Yeong C H, Satapathy S C, Molinari F and Acharya U R 2021 Feature-versus deep learning-based approaches for the automated detection of brain tumor with magnetic resonance images: A comparative study *Int. J. Imag. Syst. Technol.* https://doi.org/10.1002/ima.22646

Lightning Source UK Ltd.
Milton Keynes UK
UKHW031037260122
397711UK00005B/104